up in grasslands and brushlands. Finally, the manager must achieve sufficient control to reduce overuse in some areas while other areas go unused. Prolonged concentration of stock that results in permanent loss of vegetative cover on uplands or in riparian zones should be avoided. However, small scale loss of vegetative cover around water or feed troughs may be tolerated if surrounding vegetative cover is adequate.

Confined Livestock Production

Animal health and waste management are key issues in confined livestock operations. The moral and ethical debate taking place today regarding animal welfare is particularly intense for confined livestock production systems. The issues raised in this debate need to be addressed. Confinement livestock production is increasingly a source of surface and ground water pollutants, particularly where there are large numbers of animals per unit area. Expensive waste management facilities are now a necessary cost of confined production systems. Waste is a problem of almost all operations and must be managed with respect to both the environment and the quality of life in nearby communities. Livestock production systems that disperse stock in pastures so the wastes are not concentrated and do not overwhelm natural nutrient cycling processes have become a subject of renewed interest.

SOCIOECONOMIC ASPECTS OF SUSTAINABLE FARMING

In addition to strategies for preserving natural resources and changing production practices, sustainable agriculture requires a commitment to changing public policies, economic institutions, and social values. Strategies for change must take into account the complex, reciprocal and ever-changing relationship between agricultural production and the broader society.

The "food system" extends far beyond the farm and involves the interaction of individuals and institutions with contrasting and often competing goals including farmers, researchers, input suppliers, farmworkers, unions, farm advisors, processors, retailers, consumers, and policymakers. Relationships among these actors shift over time as new technologies spawn economic, social and political changes.A wide diversity of strategies and approaches are necessary to create a more sustainable food system. These will range from specific and concentrated efforts to alter specific policies or practices, to the longer-term tasks of reforming key institutions, rethinking economic priorities, and challenging widely-held social values. Areas of concern where change is most needed include the following:

Food and Agricultural Policy

Existing federal, state and local government policies often impede the goals of sustainable agriculture. New policies are needed to simultaneously promote environmental health, economic profitability, and social and economic equity. For example, commodity and price support programs could be restructured to allow farmers to realise the full benefits of the productivity gains made possible through alternative practices. Tax and credit policies could be modified to encourage a diverse and decentralised system of family farms rather than corporate concentration and absentee ownership. Government and land grant university research policies could be modified to emphasize the development of sustainable alternatives. Marketing orders and cosmetic standards could be amended to encourage reduced pesticide use. Coalitions must be created to address these policy concerns at the local, regional, and national level.

Land Use

Conversion of agricultural land to urban uses is a particular concern in California, as rapid growth and escalating land values threaten farming on prime soils. Existing farmland conversion patterns often discourage farmers from adopting sustainable practices and a long-term perspective on the value of land. At the same time, the close proximity of newly developed residential areas to farms is increasing the public demand for environmentally safe farming practices. Comprehensive new policies to protect prime soils and regulate development are needed, particularly in California's Central Valley. By helping farmers to adopt practices that reduce chemical use and conserve scarce resources, sustainable agriculture research and education can play a key role in building public support for agricultural land preservation. Educating land use planners and decision-makers about sustainable agriculture is an important priority.

Labor

The conditions of agricultural labor are generally far below accepted social standards and legal protections in other forms of employment. Policies and programs are needed to address this problem, working toward socially just and safe employment that provides adequate wages, working conditions, health benefits, and chances for economic stability. The needs of migrant labor for year-around employment and adequate housing are a particularly crucial problem needing immediate attention. To be more sustainable over the long-term, labor must be acknowledged and supported by government policies, recognised as important constituents of land grant universities, and carefully considered when assessing the impacts of new technologies and practices.

Rural Community Development

Rural communities are currently characterised by economic and environmental deterioration. Many are among the poorest locations in the nation. The reasons for the decline are complex, but changes in farm structure have played a significant role. Sustainable agriculture presents an opportunity to rethink the importance of family farms and rural communities. Economic development policies are needed that encourage more diversified agricultural production on family farms as a foundation for healthy economies in rural communities. In combination with other strategies, sustainable agriculture practices and policies can help foster community institutions that meet employment, educational, health, cultural and spiritual needs.

Consumers and the Food System

Consumers can play a critical role in creating a sustainable food system. Through their purchases, they send strong messages to producers, retailers and others in the system about what they think is important. Food cost and nutritional quality have always influenced consumer choices. The challenge now is to find strategies that broaden consumer perspectives, so that environmental quality, resource use, and social equity issues are also considered in shopping decisions. At the same time, new policies and institutions must be created to enable producers using sustainable practices to market their goods to a wider public. Coalitions organised around improving the food system are one specific method of creating a dialogue among consumers, retailers, producers and others. These coalitions or other public forums can be important vehicles for clarifying issues, suggesting new policies, increasing mutual trust, and encouraging a long-term view of food production, distribution and consumption.

Sustainable agriculture integrates three main goals: environmental stewardship, farm profitability, and prosperous farming communities. These goals have been defined by a variety of disciplines and may be looked at from the vantage point of the farmer or the consumer.

Sustainable agriculture refers to the ability of a farm to produce food indefinitely, without causing severe or irreversible damage to ecosystem health. Two key issues are biophysical (the long-term effects of various practices on soil properties and processes essential for crop productivity) and socio-economic (the long-term ability of farmers to obtain inputs and manage resources such as labor).

The physical aspects of sustainability are partly understood. Practices that can cause long-term damage to soil include excessive tillage and irrigation without adequate drainage (leading to salinisation). Long-term experiments have provided some of the best data on how various practices affect soil properties essential to sustainability.

Although air and sunlight are available everywhere on Earth, crops also depend on soil nutrients and the availability of water. When farmers grow and harvest crops, they remove some of these nutrients from the soil. Without replenishment, land suffers from nutrient depletion and becomes either unusable or suffers from reduced yields. Sustainable agriculture depends on replenishing the soil while minimising the use of non-renewable resources, such as natural gas, or mineral ores. Possible sources of nitrogen that would, in principle, be available indefinitely, include:

1. recycling crop waste and livestock or human manure
2. growing legume crops and forages such as peanuts or alfalfa that form symbioses with nitrogen-fixing bacteria called rhizobia
3. industrial production of nitrogen by the Haber Process uses hydrogen, which is currently derived from natural gas, (but this hydrogen could instead be made by electrolysis of water using electricity (perhaps from solar cells or windmills)) or
4. genetically engineering (non-legume) crops to form nitrogen-fixing symbioses or fix nitrogen without microbial symbionts.

The last option was proposed in the 1970s, but would be well beyond the capability of early 21st century technology, even if various concerns about biotechnology were addressed. Sustainable options for replacing other nutrient inputs (phosphorus, potassium, etc.) are more limited. An often overlooked option is landraces that are adapted to less than ideal conditions such as drought or lack of nutrients.

In some areas, sufficient rainfall is available for crop growth, but many other areas require irrigation. For irrigation systems to be sustainable they require proper management and musn't use more water from their source than is naturally replenished, otherwise the water source becomes, in effect, a non-renewable resource. Improvements in water well drilling technology and the development of submersible pumps have made it possible for large crops to be regularly grown where reliance on rainfall alone previously made this level of success unpredictable. However, this progress has come at a price, in that in many areas where this has occurred, such as the Ogallala Aquifer, the water is being used at a greater rate than its rate of recharge.

Socioeconomic aspects of sustainability are also partly understood. Regarding less concentrated farming, the best known analysis is Netting's study on smallholder systems through history.

Sustainable agriculture was also addressed by the 1990 farm bill. It was defined as follows:

- Stated by: "the term sustainable agriculture means an integrated system of plant and animal production practices having a site-specific application that will, over the long term:
    - satisfy human food and fiber needs

- enhance environmental quality and the natural resource base upon which the agricultural economy depends
- make the most efficient use of nonrenewable resources and on-farm resources and integrate, where appropriate, natural biological cycles and controls
- sustain the economic viability of farm operations
- enhance the quality of life for farmers and society as a whole."

Economics

Given the finite supply of natural resources at any specific cost and location, agriculture that is inefficient or damaging to needed resources may eventually exhaust the available resources or the ability to afford and acquire them. It may also generate negative externality, such as pollution as well as financial and production costs.

The way that crops are sold must be accounted for in the sustainability equation. Food sold locally requires little additional energy, aside from that necessary for cultivation, harvest, and transportation. Food sold at a remote location, whether at a farmers' market or the supermarket, incurs a different set of energy cost for materials, labour, and transport.

The most important factors for an individual site are sun, air, soil and water. Of the four, water and soil quality and quantity are most amenable to human intervention through time and labour.

What grows and how and where it is grown are a matter of choice. Two of the many possible practices of sustainable agriculture are crop rotation and soil amendment, both designed to ensure that crops being cultivated can obtain the necessary nutrients for healthy growth.

Methods

Many scientists, farmers, and businesses have debated how to make agriculture sustainable. One of the many practices includes growing a diverse number of perennial crops in a single field, each of which would grow in separate season so as not to compete with each other for natural resources. This system would result in increased resistance to diseases and decreased effects of erosion and loss of nutrients in soil. Nitrogen fixation from legumes, for example, used in conjunction with plants that rely on nitrate from soil for growth, helps to allow the land to be reused annually. Legumes will grow for a season and replenish the soil with ammonium and nitrate, and the next season other plants can be seeded and grown in the field in preparation for harvest.

Monoculture, a method of growing only one crop at a time in a given field, is a very widespread practice, but there are questions about its sustainability, especially if the same crop is grown every year. Growing a mixture of crops sometimes reduces disease or pest problems but polyculture has rarely, if ever, been compared to the more widespread practice of growing different crops in successive years with the same overall crop diversity; how does growing a corn-bean mixture every year compare with growing corn and bean in alternate years for example? Cropping systems that include a variety of crops may also replenish nitrogen and may also use resources such as sunlight, water, or nutrients more efficiently.

Replacing a natural ecosystem with a few specifically chosen plant varieties reduces the genetic diversity found in wildlife and makes the organisms susceptible to widespread disease. In practice, there is no single approach to sustainable agriculture, as the precise goals and methods must be adapted to each individual case. There may be some techniques of farming that are inherently in conflict with the concept of sustainability, but there is widespread misunderstanding on impacts of some practices. For example, the slash-and-burn techniques that are the characteristic feature of shifting cultivators are often cited as inherently destructive, yet slash-and-burn cultivation has been practiced in the Amazon for at least 6000 years; serious deforestation did not begin until the 1970s, largely as the result of Brazilian government programs and policies. To note that it may not have been slash-and-burn so much as slash-and-char, which with the addition of organic matter produces terra preta, one of the richest soils on Earth and the only one that regenerates itself

There are also many ways to practice sustainable animal husbandry. Some of the key tools to grazing management include fencing off the grazing area into smaller areas called paddocks, lowering stock density, and moving the stock between paddocks frequently.

Several attempts have been made to produce an artificial meat, using isolated tissues to produce it in vitro; Jason Matheny's work on this topic, whichin the New Harvest project, is one of the most commented.

Off-farm Impacts

A farm that is able to "produce perpetually", yet has negative effects on environmental quality elsewhere is not sustainable agriculture. An example of a case in which a global view may be warranted is over-application of synthetic fertiliser or animal manures, which can improve productivity of a farm but can pollute nearby rivers and coastal waters (eutrophication). The other extreme can also be undesirable, as the problem of low crop yields due to exhaustion of nutrients in the soil has been related to rainforest destruction, as in the case of slash and burn farming for livestock feed.

Sustainability affects overall production, which must increase to meet the increasing food and fiber requirements as the world's human population expands to a projected 9.3 billion people by 2050. Increased production may come from creating new farmland, which may ameliorate carbon dioxide emissions if done through reclamation of desert as in Israel, or may worsen emissions if done through slash and burn farming, as in Brazil. Additionally, genetically modified organism crops show promise for radically increasing crop yields, although many people and governments are apprehensive of this new farming method.

Some advocates of sustainable agriculture favour organic farming as the only system which can be sustained over the long-term. However, organic production methods, especially in transition, yield less than their conventional counterparts and raise the same problems of sustaining populations globally. While evidence supports organic farming during periods of drought, these figures must be interpreted with care, and modern food storage technology reduces risks associated with transient droughts. If periods of prolonged drought occur due to global warming, organic production methods can be considered as a way to adapt to a changing climate.

Urban Planning

There has been considerable debate about which form of human residential habitat may be a better social form for sustainable agriculture.

Many environmentalists pushing for increased population density to preserve agricultural land point out that urban sprawl is less sustainable and more damaging to the environment than living in the cities where cars are not needed because food and other necessities are within walking distance. However, others have theorised that sustainable ecocities, or ecovillages which combine habitation and farming with close proximity between producers and consumers, may provide greater sustainability. The use of available city space for cooperative food production is another way to achieve greater sustainability. One of the latest ideas in achieving sustainable agricultural involves shifting the production of food plants from major factory farming operations to large, urban, technical facilities called vertical farms.

The advantages of vertical farming include year-round production, isolation from pests and diseases, controllable resource recycling, and on-site production that reduces transportation costs. While a vertical farm has yet to become a reality, the idea is gaining momentum among those who believe that current sustainable farming methods will be insufficient to provide for a growing global population. For vertical farming to become a reality, billions of dollars in tax credits and subsidies will need to be made available to the operation. It may be difficult to justify spending billions of dollars on a vertical farm that will only feed 50,000 people when agriculture land remains abundant.

REFERENCES

Altieri, Miguel A. 1995. *Agroecology: The science of sustainable agriculture.* Westview Press, Boulder, CO.

Vandermeer, J. 1995. "The ecological basis of alternative agriculture". *Annual Review of Ecological Systems* 26: 201-224.

Dore, J. 1997. *Sustainability Indicators for Agriculture: Introductory Guide to Regional/National and On-farm Indicators,* Rural Industries Research and Development Corporation, Australia.

Gliessman, S.R. 1998. *Agroecology: ecological processes in sustainable agriculture.* Ann Arbor Press, Michigan.

Gold, Mary. 1999. *Sustainable Agriculture: Definitions and Terms.* Special Reference Briefs Series no. SRB 99-02 Updates SRB 94-05 September 1999. National Agricultural Library, Agricultural Research Service, U.S. Department of Agriculture.

2

Principles of Sustainable Farming

Farming sustainably means growing crops and livestock in ways that meet three objectives simultaneously:

- Economic profit
- Social benefits to the farm family and the community
- Environmental conservation

Sustainable agriculture depends on a whole-system approach whose overall goal is the continuing health of the land and people. Therefore it concentrates on long-term solutions to problems instead of short-term treatment of symptoms. Sustainable farming is more than a set of idealistic principles or a limited set of practices. Sustainability can be observed and measured; indicators that a farm or rural community is achieving the three objectives of sustainability include:

1. *Economic Sustainability*
    - The family savings or net worth is consistently going up.
    - The family debt is consistently going down.
    - The farm enterprises are consistently profitable from year to year.
    - Purchase of off-farm feed and fertiliser is decreasing.
    - Reliance on government payments is decreasing.
2. *Social Sustainability*
    - The farm supports other businesses and families in the community.
    - Dollars circulate within the local economy.
    - The number of rural families is going up or holding steady.

- Young people take over their parents' farms and continue farming.
- College graduates return to the community after graduation.

3. Environmental Sustainability
   - There is no bare ground.
   - Clean water flows in the farm's ditches and streams.
   - Wildlife is abundant.
   - Fish are prolific in streams that flow through the farm.
   - The farm landscape is diverse in vegetation.

These three objectives are managed more as a single unit, even though we must discuss them separately. The three objectives overlap constantly. Economic decisions affect the local community—buying from out of state instead of from a local supplier. Environmental decisions affect the economics—allowing soil erosion increases the need for irrigation and more fertiliser.

ENVIRONMENTAL SUSTAINABILITY

Sustainable agriculture can be viewed as ecosystem management of complex interactions among soil, water, plants, animals, climate, and people. The goal is to integrate all these factors into a production system that is appropriate for the environment, the people, and the economic conditions where the farm is located.

Farms become and stay environmentally sustainable by imitating natural systems—creating a farm landscape that mimics as closely as possible the complexity of healthy ecosystems. Nature tends to function in cycles, so that waste from one process or system becomes input for another. Industrial agriculture, in contrast, tends to function in a linear fashion similar to a factory: inputs go in one end, and products and waste come out the other. The wastes of industrial agriculture include suspended soil, nitrates, and phosphates in stream water, and nitrates and pesticides in ground water. It is a premise of sustainable agriculture that a farm is a nature-based system, not a factory.

The simpler we try to make agriculture, the more vulnerable we become to natural disasters and marketplace changes. When we try to produce a single product such as wheat, corn, or soybeans we are taking on huge risk. If instead we diversify crops and integrate plant and animal agriculture, overhead will be spread over several enterprises, reducing risk and increasing profit.

Table 1 offers some comparisons between two models of agriculture—farming as an industrial factory and farming as a biological system.

Table 1. Comparison of the industrial and biological models of agriculture.

Industrial model	*Biological model*
Energy intensive	Information intensive
Linear process	Cyclical process
Farm as factory	Farm as ecosystem
Enterprise separation	Enterprise integration
Single enterprise	Many enterprises
Monoculture	Diversity of plants and animals
Low-value products	Higher value products
Single-use equipment	Multiple-use equipment
Passive marketing	Active marketing

Farm as Ecosystem

On any farm, four major ecosystem processes are at work that, if functioning properly, will conserve the soil and water resources and eventually reduce the overall operating costs. These natural processes—energy flow, water and mineral cycles, and ecosystem dynamics—are observable and manageable.

Energy flow is the non-cyclical path of solar energy into and through any biological system. The natural world runs on sunlight. Our management decisions affect how much of it is captured and put to good use on the farm. Energy flow begins when sunlight is converted into plant growth, and continues when animals consume plants, when predator animals consume prey, and when microorganisms decompose dead plants and animals.

Some energy is lost as heat at every transfer point in the food chain. On the farm, energy capture is enhanced by maximising—both in space and in time—the leaf area available for photosynthesis, and by efficiently cycling the stored solar energy through the food chain. Off-season cover crops, perennial vegetation, and intercropping are among the tools for capturing more solar energy. Capturing sunlight and converting it to dollars is the original source of all wealth.

An effective water cycle is typified by no soil erosion, fast water entry into the soil, and the soil's capacity to store large amounts of water. Streams flow year-round from the slow release of water stored in the soil. The water cycle is improved by management decisions that add to or maintain the groundcover percentage and soil organic matter levels—the goal is to get as much water as possible into the soil during each rainfall. A surface mulch layer speeds water intake while reducing evaporation and protecting the soil from erosion. Minimising or eliminating tillage, growing high-residue crops and cover crops, and adding compost or manure to the soil maintains groundcover and builds organic matter.

Management of soil organic matter is especially important in row cropping. Raising the percentage of organic matter from 1% to 2% in sandy soil increased the available water content of that soil by 60% (from 5% of total soil volume to 8%). Such an improvement in a soil's water-holding capacity will have a beneficial effect on crop growth, especially during droughty periods.

The results of an effective water cycle are low surface runoff, low soil surface evaporation, low drought incidence, low flood incidence, high transpiration by plants, and high seepage of water to underground reservoirs.

A well-functioning mineral cycle—the movement of nutrients from the soil through the crops and animals and back to the soil—means less need for fertiliser and feed from off the farm. In nature, minerals needed for plant and animal growth are continuously recycled within the ecosystem with very little waste and no need for added fertiliser. Ultimately, to be sustainable, we need to find ways to use the natural mineral cycle to minimise our off-farm purchase of minerals.

Conditions and practices that inhibit the natural mineral cycle—erosion, nutrient leaching, organic matter depletion, selling hay or grain off the farm—tend to reduce the farm's sustainability. Practices that enhance the mineral cycle include on-farm feeding of livestock, careful management of manure and crop residues, use of catch crops to reduce nutrient leaching losses, and practices that prevent erosion. An effective ecosystem dynamic is indicated by high diversity of plants and animals both above and below ground.

"Diversity" refers not only to numbers of species, but also to genetic diversity within species, and to a broad age structure in each population. Greater diversity produces greater stability within the system and minimises pest problems. Our choices of practices and tools directly affect the level of biodiversity we have on the farm.

The first step toward increasing biodiversity on the farm is crop rotation, which helps break weed and pest life cycles and provides complementary fertilisation among the crops in the planting sequence. Advancing from rotation to strip intercrops brings a higher level of biodiversity and increases sunlight capture. Strip intercropping of corn and soybeans or cotton and alfalfa are two examples. Borders, windbreaks, and special plantings for natural enemies of pests provide habitat for beneficial organisms, further increasing biodiversity and stability.

ECONOMIC SUSTAINABILITY

Economic sustainability increasingly depends on selecting profitable enterprises, sound financial planning, proactive marketing, risk management, and good overall management. The key for row-crop producers may be to explore income opportunities other than traditional commodity crops, such as contract growing of seed corn, specialty corn, food-grade soybeans, or popcorn. These specialty crops are not for everyone; only a certain number of acres can be grown because of limited markets.

The complementary enterprises overlap with the centerpiece enterprises by sharing some of the same overhead requirements, thus lowering overall costs for all the enterprises. When we try to produce a single product such as wheat, corn, or soybeans our risk is high because "all our eggs are in one basket." When we integrate plant and animal agriculture we distribute overhead and risk among several enterprises.

Comprehensive Financial Planning

The holistic financial planning process used in Holistic Management™ provides a monthly roadmap to help people navigate through their financial year, assured that the profit will be there at year's end. The income is planned first, then a planned profit is allocated as the first expense item. The remaining expense money is allocated sequentially where it will do the most good. This sequential allocation requires that the farmer spend no more than necessary to run the enterprise for a year, while preserving the planned profit. This potent financial planning process empowers people to make decisions that are simultaneously good for the environment, the local community, and the bottom line.

Every farm needs a marketing plan of some type. Marketing can take many forms, ranging from passive marketing to the commodity chain all the way up to marketing a retail product directly to consumers. Which marketing method you choose will have a profound effect on the price your product commands. Doing some market research is essential in order to understand your market, competition, and consumer trends, and to project potential sales volume and prices. Specialty and direct markets such as organic, GMO-free, and other "green" markets yield more income but require more marketing by the producer. Direct marketing is not for everyone.

SOCIAL SUSTAINABILITY

Decisions made on the farm have effects in the local community. For example, the decision to expand your operation requires the acquisition of your neighbor's farm. To have your neighbor's farm, you must make the decision that your neighbor's farm is more important to you than your neighbor. Other examples of social decisions include: buying supplies locally rather than ordering from out of state, figuring out ways to connect local consumers with your farm, taking a consumer-oriented approach to production and management practices where both the farmer and consumer wins, and finding opportunities to ensure that neighboring communities can learn about sustainable food production.

Marketing strategies such as community supported agriculture (CSA), direct marketing through farmers' markets, school tours, and internships all have a positive impact on the local community. When people have a choice between supporting local producers or paying a little less for the products of the industrial food system, they will often choose to support their neighbours.

Farmers selling locally benefit from differentiating their products and services by qualities other than price. Fresh produce, specialty items, and locally grown and processed foods are competitive in the market place, especially when consumer education and personal contact with the farmer are part of the marketing plan. Social sustainability also includes the quality of life of those who work and live on the farm, including good communication, trust, and mutual support.

Full family participation in farm planning is an indication that the quality of life is high. Other indicators include talking openly and honestly, spending time together, a feeling of progress toward goals, and general happiness. Quality of life will be defined somewhat differently by each individual and family, based on their values and goals.

Planning and Decision-Making

Managing for three objectives simultaneously depends on clear goal-setting and effective decision-making. Several good tools for decision-making, goal-setting, and whole-farm management are available to farmers. The Kerr Center for Sustainable Agriculture, for example, has developed a sustainability checksheet with 72 criteria for quick evaluation of farming systems.

A successful transition to sustainable farming depends on the farmer's careful monitoring both of progress towards the goal and of the overall health of the system. It is useful to assume that your plan will not work, and develop a system for determining if it isn't working. For example, if the goal includes increased biodiversity, the farmer needs to know—quickly—if the grazing or cropping system being used is actually increasing the number of plant species per acre. Monitoring is particularly important in sustainable agriculture, which relies on natural systems to replace some of the work done by input products like fertiliser and pesticides.

The ability to evaluate and replan is vital to the farmer who wishes to farm more sustainably. When part of the plan is not working as intended, it becomes necessary to replan. The concept of planning- monitoring- controlling-replanning is a key characteristic of Holistic Management and is referred to as the feedback loop. The transition toward more sustainable farming requires not only planning and decision-making skills, but access to appropriate and helpful information. Fortunately, increased interest in sustainable agriculture has stimulated greater investment in research and education.

APPLYING THE PRINCIPLES

When beginning the transition, the big question is how to apply the principles of economic profitability, social enhancement, and ecological improvement in the field, in the community, and in the financial process. The decisions we make on our farms and the tools and practices we choose will determine the extent to which sustainability is realised. The ultimate goal is to farm in such a way that we extract our living as the interest, while preserving the social, water, and soil capital. We want to ensure that our activities do not compromise the landscape and community resources over the long term. Now let's look at some management concepts aimed at fostering the four ecosystem processes discussed earlier.

Strive to Heep the Soil Covered throughout the Year

Under natural conditions the soil remains covered with a skin of dead plant material, which prevents and moderates temperature extremes, increases water penetration and storage, and enhances soil aeration. Most importantly, the soil skin maintains soil structure and prevents erosion by softening the impact of falling raindrops. Bare ground, on the other hand, is vulnerable to water and wind erosion, dries out more quickly, and loses organic matter rapidly.

The major productivity costs to the farm associated with soil erosion come from the replacement of lost nutrients and reduced water holding ability, accounting for 50 to 75% of productivity loss. Soil removed by erosion typically contains about three times more nutrients than the soil left behind and is 1.5 to 5 times richer in organic matter. This organic matter loss not only results in reduced water holding capacity and degraded soil aggregation, but also loss of plant nutrients, which must then be replaced with fertilisers. Five tons of topsoil can easily contain 100 pounds of nitrogen, 60 pounds of phosphate, 45 pounds of potash, 2 pounds of calcium, 10 pounds of magnesium, and 8 pounds of sulfur. The table 2 below shows the effect of slight, moderate, and severe erosion on organic matter, soil phosphorus level, and plant-available water on a silt loam soil in Indiana.

When erosion by water and wind occurs at a rate of 7.6 tons/acre/year it costs $40/acre/year to replace the lost nutrients as fertiliser, and around $17/acre/year to pump irrigation water to replace the water holding capacity of that lost soil. The total cost of soil and water lost annually from U.S. cropland causes productivity loss of approximately $27 billion each year.

Table 2. Effect of erosion on organic matter, phosphorus, and plant-available water.

Erosion level	Organic matter	Phophorus	Plant-available water
	%	lbs./ac	%
Slight	3.0	62	7.4
Moderate	2.5	61	6.2
Severe	1.9	40	3.6

Avoid Moldboard Plowing

Soil is damaged considerably whenever it is turned over. The moldboard plow brings subsoil to the surface and buries the crop residue layer so deep it is unable to decay properly. Virtually no soil residue is left on the surface, exposing the soil to erosion and impairing the water and mineral cycles. Today, millions of acres are being farmed without any tillage at all (no-till) or in such a way that adequate groundcover remains afterwards.

Enterprise diversification

Enterprise diversification reduces financial risk by spreading income and costs (e.g., of pest control and fertiliser) out over several crops or livestock operations. Sustainability is increased when animal wastes become inputs to crop production on the same farm.

Crop Rotations

Moving from simple monoculture to a higher level of diversity begins with crop rotations, which break weed and pest life cycles, provide complementary fertilisation to crops in sequence with each other—nitrogen-fixing legume crops preceding grain crops such as corn—and prevent buildup of pest insects and weeds. In many cases, yield increases follow from the "rotation effect." Including forage crops in the rotation will reduce soil erosion and increase soil quality.

When planning crop rotations, it is important to consider that cultivated row crops—such as corn and soybeans or vegetables—tend to be soil-degrading. Since the soil is open and cultivated between rows, microbes break down organic matter at a more rapid pace. Furthermore, row crops have modest root systems and consequently do not contribute enough new organic matter to replace that lost from the open soil between rows; in most cases above-ground crop residues make only minor contributions to replacing lost organic matter.

Cereals and other crops planted with a grain drill or broadcast-seeded are more closely spaced and have more extensive root systems than row crops, greatly reducing the amount of soil exposed to degradation. In addition, they receive little or no cultivation after planting, which reduces organic-matter loss even more. As a result, cereals and green manures can be considered neutral crops, replacing soil organic matter at roughly the same rate at which it breaks down. Crops that make a perennial sod cover—such as grasses, clovers, and alfalfa—not only keep the soil entirely covered, but also have massive root systems, producing far more organic matter than is lost. Sod crops are the best soil-building crops—they can heal the damage done to soil by row cropping.

Incorporating sod crops as a fundamental part of a crop rotation not only builds soil, but supports weed-control strategies as well. Weed control improves because the types of weeds encouraged by row-cropping systems are usually not adapted to growing in a sod/hay crop.

An ideal rotation might include one year of sod crop for each year of row crop, and as many years of "neutral" crops as makes sense in the circumstances. The challenge of incorporating sod crops into a rotation is to include livestock in the system or to find a market for the hay. Sustainable production is much easier when livestock are present in the system to recycle wastes and assist in transferring nutrients from one part of the farm to another.

Fortunately, land capable of producing a 100-bushel corn yield will generally be able to produce 5-ton hay yields. With prices of $60-$70 per ton being common for ordinary hay, gross revenues per acre from hay will exceed those from corn so long as corn is under $3.00 per bushel. The net income picture is even more encouraging, however, because conventional production costs for an acre of corn are quite a bit higher than for hay.

A good crop of alfalfa fixes at least $50 worth of nitrogen every year, thus reducing fertiliser costs for the subsequent corn crop. Besides equipment costs, the major drawback to selling hay is that the nutrients it contains are shipped off the farm. Since, however, something like 75—90% of the minerals going into the front end of cattle come out the back end, keeping cattle helps retain nutrients on the farm. Cattle can serve as a very profitable method of adding value to the forage crops they consume.

Use Cover Crops and Green Manures

Perennial and biennial sod crops, annual green manures, and annual cover crops are important for building soil in field-cropping systems. Hairy vetch, for example, not only is a soil-conserving cover crop, but is capable of providing all the nitrogen required by subsequent crops like tomatoes.

The soil-building crops most appropriate for a given farm depend not only on regional factors (harshness of winter, etc.) but also on the type of production system involved: each farmer will have to determine which cover crops are most appropriate to his or her system.

Composts, Manures, and Fertilisers

Crop rotations, cover-cropping, and green-manuring are key strategies for soil building, which is the foundation of sustainable farming. However, modern production systems place high demands on land resources, requiring additional attention to soil fertility management.

Manures and composts, especially those produced on-farm or available locally at low cost, are ideal resources for cycling nutrients on-farm. From the standpoint of overall soil and crop health, composts or aged manures are preferred. Compost has a unique advantage in comparison to unaged manure and other organic soil amendments in that it has a (usually) predictable, and nearly ideal, ratio of carbon to nitrogen. Compost can be safely applied at rates of 10 tons per acre, where quantities are available. Much higher rates are not unusual, especially where soil is being improved rather than maintained.

Compost has some particular advantages in row crop production, especially when used in conjunction with cover crops and green manures. In sandy soils, compost's stable organic matter is especially effective at absorbing and retaining water. Fresh plant material incorporated as green manure, on the other hand, retains its waxy leaf coating and cannot perform the same function until thoroughly digested by microbes.

There are several conventional fertilisers that should be avoided in sustainable farming because of their harmful effects on soil organisms and structure. These include anhydrous ammonia and potassium chloride. The use of dolomite—a liming material having a high magnesium-to-calcium ratio—has also been generally discouraged, but most problems result from the frequent misuse of dolomite for raising pH on soils already high in magnesium, not from any innate detrimental qualities. It is certainly appropriate for use on fields deficient in magnesium as indicated by a proper soil test.

Some of the more "environmentally friendly" chemical fertilisers such as mono-ammonium phosphate (12-50-0), commonly called MAP, may also have a role in the transition away from the harsher chemical fertilisers. A very serviceable and affordable 4-16-16 transitional fertiliser with magnesium, sulfur, and other minor nutrients can be prepared from a combination of two-thirds sulfate of potash-magnesia and one-third mono-ammonium phosphate. When used in combination with composts and/or legume plowdowns (for nitrogen), this 4-16-16 can be banded at seeding or otherwise applied just like the regular 5-20-20, but with reduced negative impact on soil life.

Significant additions of lime, rock phosphate, and other fertilisers should be guided by soil testing to avoid soil imbalances and unnecessary expenditure on inputs. Cooperative Extension offers low-cost soil testing services in many states.

Weed Management

Weed management poses one of the greatest challenges to the crafting of sustainable production systems. However, weed populations tend to decline in severity as soil health builds. A basic understanding of weed ecology and the influence of cropping patterns on weed communities will help growers refine their use of cultural and mechanical techniques, thereby reducing the time required for effective weed control.

Prevention of weed problems is a fundamental component of management. In general terms, weed prevention in crops is based on developing a sound rotation, thwarting all attempts by existing weeds to set seed, and minimising the arrival of new weed seeds from outside the field. In a grazing system, weed management may be as simple as adding other animal species such as goats or sheep to a cattle herd to convert weeds into cash.

Certain crops can be used to smother weeds. Short-duration plantings of buckwheat and sorghum-sudangrass, for example, smother weeds by growing faster and out-competing them. In northern states, oats are commonly planted as a "nurse crop" for alfalfa, clover, and legume-grass mixtures—the oats simply take the place of weeds that would otherwise grow between the young alfalfa plants.

With enough mulch, weed numbers can be greatly reduced. Nebraska scientists applied wheat straw in early spring to a field where wheat had been harvested the previous August. At the higher straw rates, weed levels were reduced more than three times over. Wheat, like rye, is also known to possess weed-suppressing chemicals in the straw itself. This quality is known as allelopathy.

Rye is one of the most useful allelopathic cover crops because it is winter-hardy and can be grown almost anywhere. Rye residue contains generous amounts of allelopathic chemicals. When rye is killed in place and left undisturbed on the soil surface, these chemicals leach out and prevent germination of small-seeded weeds. Weed suppression is effective for about 30—60 days. If the rye is tilled into the soil, the effect is lost.

The table 3 below shows the effects of several cereal cover crops on weed production. Note that tillage alone, in the absence of any cover crop, more than doubled the number of weeds. While a good weed-prevention program will decrease weed pressure substantially, successful crop production still requires a well-conceived program for controlling weeds to the point where they have no negative impact on net income. Weed control programs include a range of carefully timed interventions designed to kill as many young seedlings as possible.

Table 3. Tillage and cover crop mulch effect on weed numbers and production.

Tillage	*Cover crop*	*Weeds/foot²*	*Weed weight pounds/foot²*
Conventional	None	12	0.22
None	None	5	0.14
None	Rye	0.9	0.1
None	Wheat	0.3	0.07
None	Barley	0.8	0.09

Insect Pest Management

Insect pests can have a serious impact on farm income. In ecologically balanced farm production systems, insect pests are always present, but massive outbreaks resulting in severe economic damage are minimised. This results in good part from the presence of natural control agents—especially predatory and parasitic insects, mites, and spiders—that keep pest populations in check. To restore populations of beneficials on the farm, cease or reduce pesticide use and other practices that harm them, and establish habitats through farmscaping.

In diverse farm systems, severe pest outbreaks are rare because natural controls exist to automatically bring populations back into balance. There is overwhelming evidence that plant mixtures (intercrops) support lower numbers of pests than pure stands. There are two schools of thought on why this occurs. One suggests that higher natural-enemy populations persist in diverse mixtures because they provide more continuous food sources (nectar, pollen, and prey) and habitat. The other thought is that pest insects who feed on only one type of plant have greater opportunity to feed, move around, and breed in pure crop stands because their resources are more concentrated than they would be in a crop mixture.

Intercropping also aids pest control efforts by reducing the ability of the pest insects to recognise their host plants. For example, thrips and white flies are attracted to green plants with a brown (soil) background, and ignore areas where vegetative cover is complete—including mulched soil. Some intercrops thus disguise the host plant from these pests by completely covering the soil. Other insects recognise their host plant by smell; onions planted with carrots mask the smell of carrots from carrot flies.

Sooner or later, nearly every grower confronts unacceptable pest pressure, making some kind of intervention necessary. Integrated pest management (IPM) is the basic framework used to decide when and how pests are controlled. The primary goal of IPM is to give growers management guidelines in order to make pest control as economically and ecologically sound as possible.

IPM integrates habitat modification and cultural, physical, biological, and chemical practices to minimise crop losses. Monitoring, record keeping, and life-cycle information about pests and their natural enemies are used to determine which control measures are needed to keep pests below an economic threshold.

Biological control—the use of living organisms to control crop pests—is one of the pillars of IPM. Biocontrol agents may be predatory, parasitic, or pathogenic; they may also be either "natural" or "applied". Biocontrol agents include insects, mites, bacteria, fungi, viruses, and nematodes. Certain beneficial nematodes (*Steinernema* species) transmit pathogens to their prey, and could be seen as a form of indirectly applied biocontrol.

When all other IPM tactics are unable to maintain insect pest populations below economic thresholds, insecticide application to control the pests and prevent economic loss is clearly justified. In such cases, farmers concerned with sustainability will usually attempt to obtain satisfactory control using one of the "biorational" pesticides, which are fairly pest-specific and usually non-persistent, causing a minimal amount of harm to beneficial organisms.

Biorational pesticides include some conventional synthetic pest control materials, but more typically are microbial insecticides like *Bacillus thuringiensis* or *Beauveria bassiana*; insecticidal soaps; pheromones, and insect growth regulators. Botanical plant extracts like neem and ryania are also known as least-toxic, narrow-spectrum controls combining minimal negative impact on beneficial species with very rapid decomposition in the environment.

Farms exploring IPM concepts for the first time may limit their involvement to monitoring levels of one or two pests on a secondary crop, applying their usual insecticide if the threshold of economic injury is approached. Others may shift from a broad-spectrum insecticide to a more beneficial-friendly material. As operator comfort with IPM increases, it is common to apply basic concepts to the primary crop and expand IPM management on the secondary crop—perhaps through the introduction of beneficial parasites or predators of the target pest insect.

As they move towards greater sustainability, IPM programs tend to go through three phases, with each stage using and building on previous knowledge and techniques:

1. The pesticide management phase, characterised by establishing economic thresholds, sampling, and spraying as needed.
2. The cultural management phase, based on a thorough understanding of the pest's biology and its relationship to the cropping system. Tactics employed to control pests include delayed planting dates, crop rotation, altered harvest dates, etc.

3. The biological control phase, or "bio-intensive IPM," requires thorough understanding of the biology of natural enemies and an ability to measure how effective these agents are in controlling pests. When natural agents do not meet expected goals, the IPM practitioner uses "soft" pesticides and applications are timed to minimise pesticide exposure of beneficials.

Plant Disease Management

The first step toward preventing serious disease problems in any cropping system is the production of healthy plants nurtured by a microbially active soil. Healthy soil suppresses root diseases naturally; the primary means to create disease-suppressive soil is to add biologically active compost at appropriate rates to a soil with balanced mineral levels. Supplemental strategies include crop rotation, resistant cultivars, good soil drainage, adequate air movement, and planting clean seed.

Biorational fungicides include compost teas baking soda, and plant extracts. As with insect pest management, integrated management principles should be applied, including monitoring of environmental conditions, to determine whether preventive fungicidal sprays are required.

REFERENCES

Horne, James E. and Maura McDermott. 2001. *The Next Green Revolution: Essential Steps to a Healthy, Sustainable Agriculture.* Food Products Press, an imprint of The Haworth Press, Binghamton, NY. 312 p.

Pretty, J.N. 1994. *Regenerating Agriculture.* Earthscan Publications Ltd., London.

Reijntjes, C.B., Haverkort and A. Waters-Bayer. 1992. *Farming for the future.* MacMillan Press Ltd., London.

Sustainable Agriculture Network. 2002. *Resources from the Sustainable Agriculture Network.* Sustainable Agriculture Research and Education (SARE) Program. Sustainable Agriculture Publications, 210 UVM, Hills Building, Burlington, VT 05405-0082.

Suthipradit, S. et.al. (eds) "Sustainable Agriculture: Possibility and Direction" Proceedings of the 2nd Asia-Pacific Conference on Sustainable Agriculture 18-20 October 1999, Phitsanulok, Thailand.

3

Sustainable Soil Management

A common philosophy among sustainable agriculture practitioners is that a "healthy" soil is a key component of sustainability; that is, a healthy soil will produce healthy crop plants that have optimum vigor and are less susceptible to pests. While many crops have key pests that attack even the healthiest of plants, proper soil, water and nutrient management can help prevent some pest problems brought on by crop stress or nutrient imbalance. Furthermore, crop management systems that impair soil quality often result in greater inputs of water, nutrients, pesticides, and/or energy for tillage to maintain yields.

In sustainable systems, the soil is viewed as a fragile and living medium that must be protected and nurtured to ensure its long-term productivity and stability. Methods to protect and enhance the productivity of the soil include using cover crops, compost and/or manures, reducing tillage, avoiding traffic on wet soils, and maintaining soil cover with plants and/or mulches. Conditions in most soils (warm, irrigated, and tilled) do not favor the buildup of organic matter. Regular additions of organic matter or the use of cover crops can increase soil aggregate stability, soil tilth, and diversity of soil microbial life.

Any farmer will tell you that a good soil:

- feels soft and crumbles easily
- drains well and warms up quickly in the spring
- does not crust after planting
- soaks up heavy rains with little runoff
- stores moisture for drought periods
- has few clods and no hardpan
- resists erosion and nutrient loss

- supports high populations of soil organisms
- has a rich, earthy smell
- does not require increasing inputs for high yields
- produces healthy, high-quality crops

All these criteria indicate a soil that functions effectively today and will continue to produce crops long into the future. These characteristics can be created through management practices that optimise the processes found in native soils. How does soil in its native condition function? How do forests and native grasslands produce plants and animals in the complete absence of fertiliser and tillage? Understanding the principles by which native soils function can help farmers develop and maintain productive and profitable soil both now and for future generations.

The soil, the environment, and farm condition benefit when the soil's natural productivity is managed in a sustainable way. Reliance on purchased inputs declines year by year, while land value and income potential increase. Some of the things we spend money on can be done by the natural process itself for little or nothing. Good soil management produces crops and animals that are healthier, less susceptible to disease, and more productive. To understand this better, let's start with the basics.

SOIL: TEXTURE AND STRUCTURE

Soils are made up of four basic components: minerals, air, water, and organic matter. In most soils, minerals represent around 45% of the total volume, water and air about 25% each, and organic matter from 2% to 5%. The mineral portion consists of three distinct particle sizes classified as sand, silt, or clay. Sand is the largest particle that can be considered soil. Sand is largely the mineral quartz, though other minerals are also present. Quartz contains no plant nutrients, and sand cannot hold nutrients—they leach out easily with rainfall. Silt particles are much smaller than sand, but like sand, silt is mostly quartz.

The smallest of all the soil particles is clay. Clays are quite different from sand or silt, and most types of clay contain appreciable amounts of plant nutrients. Clay has a large surface area resulting from the plate-like shape of the individual particles. Sandy soils are less productive than silts, while soils containing clay are the most productive and use fertilisers most effectively. Soil texture refers to the relative proportions of sand, silt, and clay.

A loam soil contains these three types of soil particles in roughly equal proportions. A sandy loam is a mixture containing a larger amount of sand and a smaller amount of clay, while a clay loam contains a larger amount of clay and a smaller amount of sand. These and other texture designations are listed in Table 1.

Another soil characteristic—soil structure—is distinct from soil texture. Structure refers to the clumping together or "aggregation" of sand, silt, and clay particles into larger secondary clusters. If you grab a handful of soil, good structure is apparent when the soil crumbles easily in your hand. This is an indication that the sand, silt, and clay particles are aggregated into granules or crumbs.

Both texture and structure determine pore space for air and water circulation, erosion resistance, looseness, ease of tillage, and root penetration. While texture is related to the minerals in the soil and does not change with agricultural activities, structure can be improved or destroyed readily by choice and timing of farm practices.

Table 1. Soil texture designations ranging from coarse to fine.

Texture	*Designation*
Coarse-textured	Sand
	Loamy Sand
	Sandy loam
	Fine sandy loam
	Loam
	Silty loam
	Silt
	Silty clay loam
	Clay loam
Fine-textured	Clay

IMPORTANCE OF SOIL ORGANISMS

An acre of living topsoil contains approximately 900 pounds of earthworms, 2,400 pounds of fungi, 1,500 pounds of bacteria, 133 pounds of protozoa, 890 pounds of arthropods and algae, and even small mammals in some cases. Therefore, the soil can be viewed as a living community rather than an inert body.

Soil organic matter also contains dead organisms, plant matter, and other organic materials in various phases of decomposition. Humus, the dark-colored organic material in the final stages of decomposition, is relatively stable. Both organic matter and humus serve as reservoirs of plant nutrients; they also help to build soil structure and provide other benefits. The type of healthy living soil required to support humans now and far into the future will be balanced in nutrients and high in humus, with a broad diversity of soil organisms. It will produce healthy plants with minimal weed, disease, and insect pressure. To accomplish this, we need to work with the natural processes and optimise their functions to sustain our farms.

Considering the natural landscape, you might wonder how native prairies and forests function in the absence of tillage and fertilisers. These soils are tilled by soil organisms, not by machinery. They are fertilised too, but the fertility is used again and again and never leaves the site. Native soils are covered with a layer of plant litter and/or growing plants throughout the year. Beneath the surface litter, a rich complexity of soil organisms decompose plant residue and dead roots, then release their stored nutrients slowly over time. In fact, topsoil is the most biologically diverse part of the earth. Soil-dwelling organisms release bound-up minerals, converting them into plant-available forms that are then taken up by the plants growing on the site.

The organisms recycle nutrients again and again with the death and decay of each new generation of plants. There are many different types of creatures that live on or in the topsoil. Each has a role to play. These organisms will work for the farmer's benefit if we simply manage for their survival. Consequently, we may refer to them as soil livestock. While a great variety of organisms contribute to soil fertility, earthworms, arthropods, and the various microorganisms merit particular attention.

Earthworms

Earthworm burrows enhance water infiltration and soil aeration. Fields that are "tilled" by earthworm tunneling can absorb water at a rate 4 to 10 times that of fields lacking worm tunnels. This reduces water runoff, recharges groundwater, and helps store more soil water for dry spells. Vertical earthworm burrows pipe air deeper into the soil, stimulating microbial nutrient cycling at those deeper levels. When earthworms are present in high numbers, the tillage provided by their burrows can replace some expensive tillage work done by machinery.

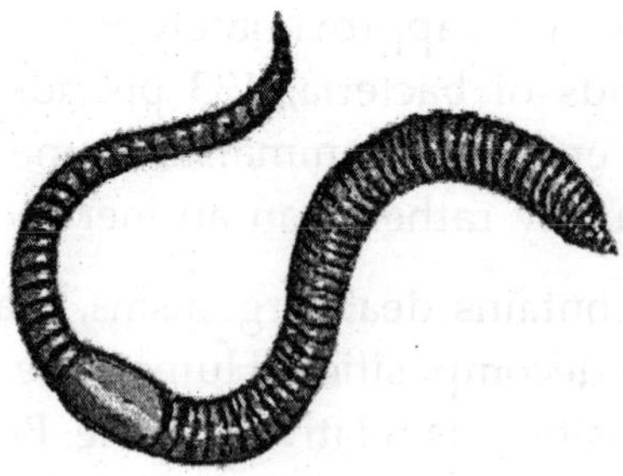

Figure 1. Earthworm

Worms eat dead plant material left on top of the soil and redistribute the organic matter and nutrients throughout the topsoil layer. Nutrient-rich organic compounds line their tunnels, which may remain in place for years if not disturbed. During droughts these tunnels allow for deep plant root penetration into subsoil regions of higher moisture content. In addition to organic matter, worms also consume soil and soil microbes. The soil clusters they expel from their digestive tracts are known as *worm casts* or *castings*. These range from the size of a mustard seed to that of a sorghum seed, depending on the size of the worm.

The soluble nutrient content of worm casts is considerably higher than that of the original soil. A good population of earthworms can process 20,000 pounds of topsoil per year—with turnover rates as high as 200 tons per acre having been reported in some exceptional cases. Earthworms also secrete a plant growth stimulant. Reported increases in plant growth following earthworm activity may be partially attributed to this substance, not just to improved soil quality.

Table 2. Selected nutrient analyses of worm casts compared to those of the surrounding soil.

Nutrient	*Worm casts (Lbs/ac)*	*Soil (Lbs/ac)*
Carbon	171,000	78,500
Nitrogen	10,720	7,000
Phosphorus	280	40
Potassium	900	140

Earthworms thrive where there is no tillage. Generally, the less tillage the better, and the shallower the tillage the better. Worm numbers can be reduced by as much as 90% by deep and frequent tillage. Tillage reduces earthworm populations by drying the soil, burying the plant residue they feed on, and making the soil more likely to freeze. Tillage also destroys vertical worm burrows and can kill and cut up the worms themselves. Worms are dormant in the hot part of the summer and in the cold of winter. Young worms emerge in spring and fall-they are most active just when farmers are likely to be tilling the soil. Table 3 shows the effect of tillage and cropping practices on earthworm numbers.

Table 3. Effect of crop management on earthworm populations.

Crop	*Management*	*Worms/foot*^{2}
Corn	Plow	1
Corn	No-till	2
Soybean	Plow	6
Soybean	No-till	14
Bluegrass/clover	--	39
Dairy pasture	--	33

As a rule, earthworm numbers can be increased by reducing or eliminating tillage (especially fall tillage), not using a moldboard plow, reducing residue particle size (using a straw chopper on the combine), adding animal manure, and growing green manure crops. It is beneficial to leave as much surface residue as possible year-round. Cropping systems that typically have the most earthworms are perennial cool-season grass grazed rotationally, warm-season perennial grass grazed rotationally, and annual croplands using no-till. Ridge-till and strip tillage will generally have more earthworms than clean tillage involving plowing and disking. Cool season grass rotationally grazed is highest because it provides an undisturbed environment plus abundant organic matter from the grass roots and fallen grass litter. Generally speaking, worms want their food on top, and they want to be left alone. Earthworms prefer a near-neutral soil pH, moist soil conditions, and plenty of plant residue on the soil surface. They are sensitive to certain pesticides and some incorporated fertilisers.

Some insecticides in the organophosphate family are mildly toxic to earthworms, while synthetic pyrethroids are harmless to them. Most herbicides have little effect on worms except for the triazines, such as Atrazine, which are moderately toxic. Also, anhydrous ammonia kills earthworms in the injection zone because it dries the soil and temporarily increases the pH there. High rates of ammonium-based fertilisers are also harmful.

Arthropods

In addition to earthworms, there are many other species of soil organisms that can be seen by the naked eye. Among them are sowbugs, millipedes, centipedes, slugs, snails, and springtails. These are the primary decomposers. Their role is to eat and shred the large particles of plant and animal residues. Some bury residue, bringing it into contact with other soil organisms that further decompose it.

Some members of this group prey on smaller soil organisms. The springtails are small insects that eat mostly fungi. Their waste is rich in plant nutrients released after other fungi and bacteria decompose it. Also of interest are dung beetles, which play a valuable role in recycling manure and reducing livestock intestinal parasites and flies.

Bacteria

Bacteria are the most numerous type of soil organism: every gram of soil contains at least a million of these tiny one-celled organisms. There are many different species of bacteria, each with its own role in the soil environment. One of the major benefits bacteria provide for plants is in making nutrients available to them. Some species release nitrogen, sulfur, phosphorus, and trace elements from organic matter. Others break down soil minerals, releasing potassium, phosphorus, magnesium, calcium, and iron. Still other species make and release plant growth hormones, which stimulate root growth.

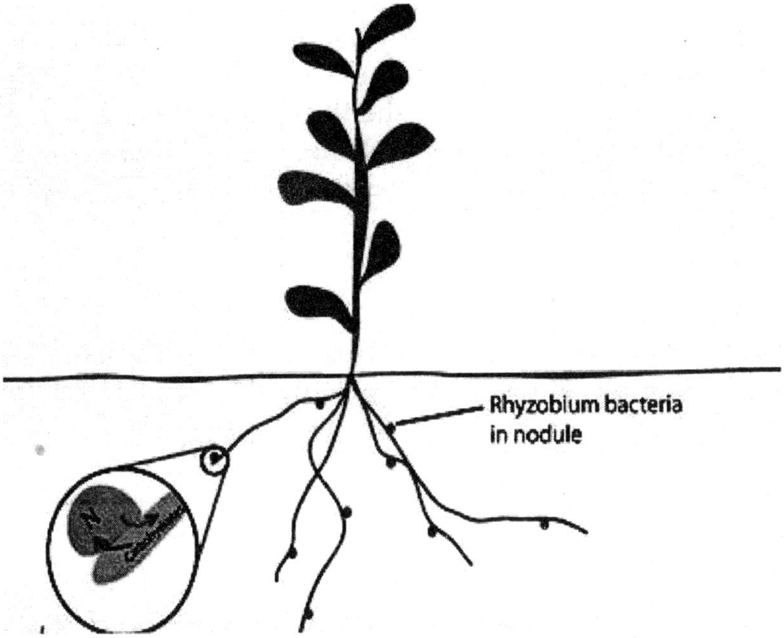

Figure 2. Function of Nitrogen fixing bacteria

Several species of bacteria transform nitrogen from a gas in the air to forms available for plant use, and from these forms back to a gas again. A few species of bacteria fix nitrogen in the roots of legumes, while others fix nitrogen independently of plant association. Bacteria are responsible for converting nitrogen from ammonium to nitrate and back again, depending on certain soil conditions. Other benefits to plants provided by various species of bacteria include increasing the solubility of nutrients, improving soil structure, fighting root diseases, and detoxifying soil.

Fungi

Fungi come in many different species, sizes, and shapes in soil. Some species appear as thread-like colonies, while others are one-celled yeasts. Slime molds and mushrooms are also fungi. Many fungi aid plants by breaking down organic matter or by releasing nutrients from soil minerals. Fungi are generally quick to colonise larger pieces of organic matter and begin the decomposition process. Some fungi produce plant hormones, while others produce antibiotics including penicillin. There are even species of fungi that trap harmful plant-parasitic nematodes.

The mycorrhizae are fungi that live either on or in plant roots and act to extend the reach of root hairs into the soil. Mycorrhizae increase the uptake of water and nutrients, especially phosphorus. They are particularly important in degraded or less fertile soils. Roots colonised by mycorrhizae are less likely to be penetrated by root-feeding nematodes, since the pest cannot pierce the thick fungal network. Mycorrhizae also produce hormones and antibiotics that enhance root growth and provide disease suppression. The fungi benefit by taking nutrients and carbohydrates from the plant roots they live in.

Actinomycetes

Actinomycetes are thread-like bacteria that look like fungi. While not as numerous as bacteria, they too perform vital roles in the soil. Like the bacteria, they help decompose organic matter into humus, releasing nutrients. They also produce antibiotics to fight diseases of roots. Many of these same antibiotics are used to treat human diseases. Actinomycetes are responsible for the sweet, earthy smell noticed whenever a biologically active soil is tilled.

Algae

Many different species of algae live in the upper half-inch of the soil. Unlike most other soil organisms, algae produce their own food through photosynthesis. They appear as a greenish film on the soil surface following a saturating rain. Algae improve soil structure by producing slimy substances that glue soil together into water-stable aggregates. Some species of algae (the blue-greens) can fix their own nitrogen, some of which is later released to plant roots.

Protozoa

Protozoa are free-living microorganisms that crawl or swim in the water between soil particles. Many soil protozoa are predatory, eating other microbes. One of the most common is an amoeba that eats bacteria. By eating and digesting bacteria, protozoa speed up the cycling of nitrogen from the bacteria, making it more available to plants.

Nematodes

Nematodes are abundant in most soils, and only a few species are harmful to plants. The harmless species eat decaying plant litter, bacteria, fungi, algae, protozoa, and other nematodes. Like other soil predators, nematodes speed the rate of nutrient cycling.

Soil Organisms and Soil Quality

All these organisms—from the tiny bacteria up to the large earthworms and insects—

interact with one another in a multitude of ways in the soil ecosystem. Organisms not directly involved in decomposing plant wastes may feed on each other or each other's waste products or the other substances they release. Among the substances released by the various microbes are vitamins, amino acids, sugars, antibiotics, gums, and waxes. Roots can also release into the soil various substances that stimulate soil microbes. These substances serve as food for select organisms. Some scientists and practitioners theorize that plants use this means to stimulate the specific population of microorganisms capable of releasing or otherwise producing the kind of nutrition needed by the plants. Table 4 for estimates of typical amounts of various organisms found in fertile soil.

Table 4. Weights of soil organisms in the top 7 inches of fertile soil.

Organism	*Pounds of liveweight/acre*
Bacteria	1000
Actinomycetes	1000
Molds	2000
Algae	100
Protozoa	200
Nematodes	50
Insects	100
Worms	1000
Plant roots	2000

ORGANIC MATTER AND SOIL FOODWEB

Understanding the role that soil organisms play is critical to sustainable soil management. Based on that understanding, focus can be directed toward strategies that build both the numbers and the diversity of soil organisms. Like cattle and other farm animals, soil livestock require proper feed. That feed comes in the form of organic matter.

Organic matter and humus are terms that describe somewhat different but related things. Organic matter refers to the fraction of the soil that is composed of both living organisms and once-living residues in various stages of decomposition. Humus is only a small portion of the organic matter. It is the end product of organic matter decomposition and is relatively stable. Further decomposition of humus occurs very slowly in both agricultural and natural settings. In natural systems, a balance is reached between the amount of humus formation and the amount of humus decay. This balance also occurs in most agricultural soils, but often at a much lower level of soil humus. Humus contributes to well-structured soil that, in turn, produces high-quality plants. It is clear that management of organic matter and humus is essential to sustaining the whole soil ecosystem.

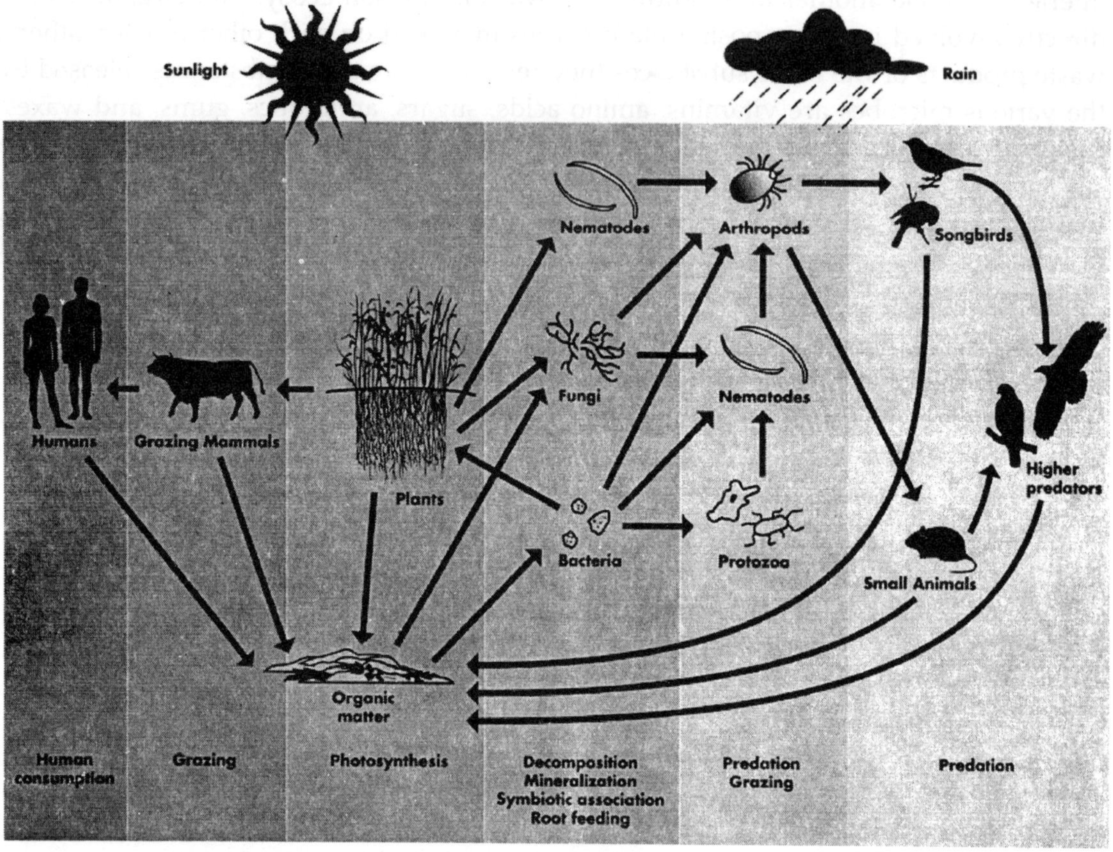

Figure 3. Soilfood web

The benefits of a topsoil rich in organic matter and humus are many. They include rapid decomposition of crop residues, granulation of soil into water-stable aggregates, decreased crusting and clodding, improved internal drainage, better water infiltration, and increased water and nutrient holding capacity. Improvements in the soil's physical structure facilitate easier tillage, increased water storage capacity, reduced erosion, better formation and harvesting of root crops, and deeper, more prolific plant root systems.

Soil organic matter can be compared to a bank account for plant nutrients. Soil containing 4% organic matter in the top seven inches has 80,000 pounds of organic matter per acre. That 80,000 pounds of organic matter will contain about 5.25% nitrogen, amounting to 4,200 pounds of nitrogen per acre. Assuming a 5% release rate during the growing season, the organic matter could supply 210 pounds of nitrogen to a crop. However, if the organic matter is allowed to degrade and lose nitrogen, purchased fertiliser will be necessary to prop up crop yields.

All the soil organisms mentioned previously, except algae, depend on organic matter as their food source. Therefore, to maintain their populations, organic matter must be renewed from plants growing on the soil, or from animal manure, compost, or other materials imported from off site. When soil livestock are fed, fertility is built up in the soil, and the soil will feed the plants.

Ultimately, building organic matter and humus levels in the soil is a matter of managing the soil's living organisms—something akin to wildlife management or animal husbandry. This entails working to maintain favorable conditions of moisture, temperature, nutrients, pH, and aeration. It also involves providing a steady food source of raw organic material.

Soil Tilth and Organic Matter

A soil that drains well, does not crust, takes in water rapidly, and does not make clods is said to have good tilth. Tilth is the physical condition of the soil as it relates to tillage ease, seedbed quality, easy seedling emergence, and deep root penetration. Good tilth is dependent on aggregation—the process whereby individual soil particles are joined into clusters or "aggregates."

Aggregates form in soils when individual soil particles are oriented and brought together through the physical forces of wetting and drying or freezing and thawing. Weak electrical forces from calcium and magnesium hold soil particles together when the soil dries. When these aggregates become wet again, however, their stability is challenged, and they may break apart. Aggregates can also be held together by plant roots, earthworm activity, and by glue-like products produced by soil microorganisms. Earthworm-created aggregates are stable once they come out of the worm. An aggregate formed by physical forces can be bound together by fine root hairs or threads produced by fungi.

Crusting is a common problem on soils that are poorly aggregated. Crusting results chiefly from the impact of falling raindrops. Rainfall causes clay particles on the soil surface to disperse and clog the pores immediately beneath the surface. Following drying, a sealed soil surface results in which most of the pore space has been drastically reduced due to clogging from dispersed clay particles. Subsequent rainfall is much more likely to run off than to flow into the soil.

Since raindrops start crusting, any management practices that protect the soil from their impact will decrease crusting and increase water flow into the soil. Mulches and cover crops serve this purpose well, as do no-till practices, which allow the accumulation of surface residue. Also, a well-aggregated soil will resist crusting because the water-stable aggregates are less likely to break apart when a raindrop hits them.

Long-term grass production produces the best-aggregated soils. A grass sod extends a mass of fine roots throughout the topsoil, contributing to the physical processes that help form aggregates. Roots continually remove water from soil microsites, providing local wetting and drying effects that promote aggregation. Fine root hairs also bind soil aggregates together. Roots also produce food for soil microorganisms and earthworms, which in turn generate compounds that bind soil particles into water-stable aggregates. In addition, perennial grass sods provide protection from raindrops and erosion. Thus, a perennial cover creates a combination of conditions optimal for the creation and maintenance of well-aggregated soil.

Conversely, cropping sequences that involve annual plants and extensive cultivation provide less vegetative cover and organic matter, and usually result in a rapid decline in soil aggregation.

Farming practices can be geared to conserve and promote soil aggregation. Because the binding substances are themselves susceptible to microbial degradation, organic matter needs to be replenished to maintain microbial populations and overall aggregated soil status. Practices should conserve aggregates once they are formed, by minimising factors that degrade and destroy aggregation. Some factors that destroy or degrade soil aggregates are:

- bare soil surface exposed to the impact of raindrops
- removal of organic matter through crop production and harvest without return of organic matter to the soil
- excessive tillage
- working the soil when it is too wet or too dry
- use of anhydrous ammonia, which speeds up decomposition of organic matter
- excess nitrogen fertilisation
- allowing the build-up of excess sodium from irrigation or sodium-containing fertilisers

Tillage, Organic Matter, and Plant Productivity

Several factors affect the level of organic matter that can be maintained in a soil. Among these are organic matter additions, moisture, temperature, tillage, nitrogen levels, cropping, and fertilisation. The level of organic matter present in the soil is a direct function of how much organic material is being produced or added to the soil versus the rate of decomposition. Achieving this balance entails slowing the speed of organic matter decomposition, while increasing the supply of organic materials produced on site and/or added from off site.

Moisture and temperature also profoundly affect soil organic matter levels. High rainfall and temperature promote rapid plant growth, but these conditions are also favorable to rapid organic matter decomposition and loss. Low rainfall or low temperatures slow both plant growth and organic matter decomposition. The native Midwest prairie soils originally had a high amount of organic matter from the continuous growth and decomposition of perennial grasses, combined with a moderate temperature that did not allow for rapid decomposition of organic matter. Moist and hot tropical areas may appear lush because of rapid plant growth, but soils in these areas are low in nutrients. Rapid decomposition of organic matter returns nutrients back to the soil, where they are almost immediately taken up by rapidly growing plants.

Tillage can be beneficial or harmful to a biologically active soil, depending on what type of tillage is used and when it is done. Tillage affects both erosion rates and soil organic matter decomposition rates. Tillage can reduce the organic matter level in croplands below 1%, rendering them biologically dead. Clean tillage involving moldboard plowing and disking breaks down soil aggregates and leaves the soil prone to erosion from wind and water. The moldboard plow can bury crop residue and topsoil to a depth of 14 inches.

At this depth, the oxygen level in the soil is so low that decomposition cannot proceed adequately. Surface-dwelling decomposer organisms suddenly find themselves suffocated and soon die. Crop residues that were originally on the surface but now have been turned under will putrefy in the oxygen-deprived zone. This rotting activity may give a putrid smell to the soil. Furthermore, the top few inches of the field are now often covered with subsoil having very little organic matter content and, therefore, limited ability to support productive crop growth.

The topsoil is where the biological activity happens—it's where the oxygen is. That's why a fence post rots off at the surface. In terms of organic matter, tillage is similar to opening the air vents on a wood-burning stove; adding organic matter is like adding wood to the stove. Ideally, organic matter decomposition should proceed as an efficient burn of the "wood" to release nutrients and carbohydrates to the soil organisms and create stable humus. Shallow tillage incorporates residue and speeds the decomposition of organic matter by adding oxygen that microbes need to become more active.

In cold climates with a long dormant season, light tillage of a heavy residue may be beneficial; in warmer climates it is hard enough to maintain organic matter levels without any tillage.

Tillage also reduces the rate of water entry into the soil by removal of ground cover and destruction of aggregates, resulting in compaction and crusting. Table 5 shows three different tillage methods and how they affect water entry into the soil. Notice the direct relationship between tillage type, ground cover, and water infiltration. No-till has more than three times the water infiltration of the moldboard-plowed soil. Additionally, no-till fields will have higher aggregation from the organic matter decomposition on site. The surface mulch typical of no-till fields acts as a protective skin for the soil. This soil skin reduces the impact of raindrops and buffers the soil from temperature extremes as well as reducing water evaporation.

Table 5. Tillage effects on water infiltration and ground cover.

	Water Infiltration mm/minute	*Ground Cover Percent*
No-till	2.7	48
Chisel Plow	1.3	27
Moldboard Plow	0.8	12

Both no-till and reduced-tillage systems provide benefits to the soil. The advantages of a no-till system include superior soil conservation, moisture conservation, reduced water runoff, long-term buildup of organic matter, and increased water infiltration. A soil managed without tillage relies on soil organisms to take over the job of plant residue incorporation formerly done by tillage. On the down side, no-till can foster a reliance on herbicides to control weeds and can lead to soil compaction from the traffic of heavy equipment.

Pioneering development work on chemical-free no-till farming is proceeding at several research stations and farms in the eastern U.S. Pennsylvania farmer Steve Groff has been farming no-till with minimal or no herbicides for several years. Groff grows cover crops extensively in his fields, rolling them down in the spring using a 10-foot rolling stalk chopper. This rolling chopper kills the rye or vetch cover crop and creates a nice no-till mulch into which he plants a variety of vegetable and grain crops. After several years of no-till production, his soils are mellow and easy to plant into.

Other conservation tillage systems include ridge tillage, minimum tillage, zone tillage, and reduced tillage, each possessing some of the advantages of both conventional till and no-till. These systems represent intermediate tillage systems, allowing more flexibility than either a no-till or conventional till system might. They are more beneficial to soil organisms than a conventional clean-tillage system of moldboard plowing and disking.

Adding manure and compost is a recognised means for improving soil organic matter and humus levels. In their absence, perennial grass is the only crop that can regenerate and increase soil humus.

Cool-season grasses build soil organic matter faster than warm-season grasses because they are growing much longer during a given year. When the soil is warm enough for soil organisms to decompose organic matter, cool-season grass is growing. While growing, it is producing organic matter and cycling minerals from the decomposing organic matter in the soil. In other words, there is a net gain of organic matter because the cool-season grass is producing organic matter faster than it is being used up.

With warm-season grasses, organic matter production during the growing season can be slowed during the long dormant season from fall through early spring. During the beginning and end of this dormant period, the soil is still biologically active, yet no grass growth is proceeding. Some net accumulation of organic matter can occur under warm-season grasses, however. In a Texas study, switchgrass grown for four years increased soil carbon content from 1.1% to 1.5% in the top 12 inches of soil. In hot and moist regions, a cropping rotation that includes several years of pasture will be most beneficial.

Effect of Nitrogen on Organic Matter

Excessive nitrogen applications stimulate increased microbial activity, which in turn speeds organic matter decomposition. The extra nitrogen narrows the ratio of carbon to nitrogen in the soil. Native or uncultivated soils have approximately 12 parts of carbon to each part of nitrogen, or a C:N ratio of 12:1. At this ratio, populations of decay bacteria are kept at a stable level, since additional growth in their population is limited by a lack of nitrogen. When large amounts of inorganic nitrogen are added, the C:N ratio is reduced, which allows the populations of decay organisms to explode as they decompose more organic matter with the now abundant nitrogen.

While soil bacteria can efficiently use moderate applications of inorganic nitrogen accompanied by organic amendments (carbon), excess nitrogen results in decomposition of existing organic matter at a rapid rate. Eventually, soil carbon content may be reduced to a level where the bacterial populations are on a starvation diet. With little carbon available, bacterial populations shrink, and less of the free soil nitrogen is absorbed. Thereafter, applied nitrogen, rather than being cycled through microbial organisms and re-released to plants slowly over time, becomes subject to leaching. This can greatly reduce the efficiency of fertilisation and lead to environmental problems.

To minimise the fast decomposition of soil organic matter, carbon should be added with nitrogen. Typical carbon sources—such as green manures, animal manure, and compost—serve this purpose well.

Amendments containing too high a carbon to nitrogen ratio (25:1 or more) can tip the balance the other way, resulting in nitrogen being tied up in an unavailable form.

Soil organisms consume all the nitrogen in an effort to decompose the abundant carbon; tied up in the soil organisms, nitrogen remains unavailable for plant uptake. As soon as a soil microorganism dies and decomposes, its nitrogen is consumed by another soil organism, until the balance between carbon and nitrogen is achieved again.

CONVENTIONAL FERTILISERS

Commercial fertiliser can be a valuable resource to farmers in transition to a more sustainable system and can help meet nutrient needs during times of high crop nutrient demand or when weather conditions result in slow nutrient release from organic resources. Commercial fertilisers have the advantage of supplying plants with immediately available forms of nutrients. They are often less expensive and less bulky to apply than many natural fertilisers.

Not all conventional fertilisers are alike. Many appear harmless to soil livestock, but some are not. Anhydrous ammonia contains approximately 82% nitrogen and is applied subsurface as a gas. Anhydrous speeds the decomposition of organic matter in the soil, leaving the soil more compact as a result. The addition of anhydrous causes increased acidity in the soil, requiring 148 pounds of lime to neutralise 100 pounds of anhydrous ammonia, or 1.8 pounds of lime for every pound of nitrogen contained in the anhydrous.

Anhydrous ammonia initially kills many soil microorganisms in the application zone. Bacteria and actinomycetes recover within one to two weeks to levels higher than those prior to treatment. Soil fungi, however, may take seven weeks to recover. During the recovery time, bacteria are stimulated to grow more, and decompose more organic matter, by the high soil nitrogen content.

As a result, their numbers increase after anhydrous applications, then decline as available soil organic matter is depleted. Farmers commonly report that the long-term use of synthetic fertilisers, especially anhydrous ammonia, leads to soil compaction and poor tilth. When bacterial populations and soil organic matter decrease, aggregation declines, because existing glues that stick soil particles together are degraded, and no other glues are being produced.

Potassium chloride (KCl) (0-0-60 and 0-0-50), also known as muriate of potash, contains approximately 50 to 60% potassium and 47.5% chloride. Muriate of potash is made by refining potassium chloride ore, which is a mixture of potassium and sodium salts and clay from the brines of dying lakes and seas. The potential harmful effects from KCl can be surmised from the salt concentration of the material. Table 6 shows that, pound for pound, KCl is surpassed only by table salt on the salt index. Additionally, some plants such as tobacco, potatoes, peaches, and some legumes are especially sensitive to chloride. High rates of KCl must be avoided on such crops. Potassium sulfate, potassium nitrate, sul-po-mag, or organic sources of potassium may be considered as alternatives to KCl for fertilisation.

Table 6. Salt index for various fertilisers.

Material	*Salt Index*	*Salt index per unit of plant food*
Sodium chloride	153	2.90
Potassium chloride	116	1.90
Ammonium nitrate	105	3.00
Sodium nitrate	100	6.10
Urea	75	1.60
Potassium nitrate	74	1.60
Ammonium sulfate	69	3.30
Calcium nitrate	53	4.40
Anhydrous ammonia	47	0.06
Sulfate-potash-magnesia	43	2.00
Di-ammonium phosphate	34	1.60
Monammonium phosphate	30	2.50
Gypsum	8	0.03
Calcium carbonate	5	0.01

Sodium nitrate, also known as Chilean nitrate or nitrate of soda, is another high-salt fertiliser. Because of the relatively low nitrogen content of sodium nitrate, a high amount of sodium is added to the soil when normal applications of nitrogen are made with this material. The concern is that excessive sodium acts as a dispersant of soil particles, degrading aggregation. The salt index for KCl and sodium nitrate can be seen in Table 6.

PROPERTIES OF TOPSOIL

Topsoil is the capital reserve of every farm. Ever since mankind started agriculture, erosion of topsoil has been the single largest threat to a soil's productivity—and, consequently, to farm profitability. This is still true today. In the U.S., the average acre of cropland is eroding at a rate of 7 tons per year. To sustain agriculture means to sustain soil resources, because that's the source of a farmer's livelihood.

The major productivity costs to the farm associated with soil erosion come from the replacement of lost nutrients and reduced water holding ability, accounting for 50 to 75% of productivity loss. Soil that is removed by erosion typically contains about three times more nutrients than the soil left behind and is 1.5 to 5 times richer in organic matter. This organic matter loss not only results in reduced water holding capacity and degraded soil aggregation, but also loss of plant nutrients, which must then be replaced with nutrient amendments.

Five tons of topsoil (the so-called tolerance level) can easily contain 100 pounds of nitrogen, 60 pounds of phosphate, 45 pounds of potash, 2 pounds of calcium, 10 pounds of magnesium, and 8 pounds of sulfur.

When erosion by water and wind occurs at a rate of 7.6 tons/acre/year it costs $40 per acre each year to replace the lost nutrients as fertiliser and around $17/acre/year to pump well irrigation water to replace the soil water holding capacity of that lost soil.

Water erosion gets started when falling rainwater collides with bare ground and detaches soil particles from the parent soil body. After enough water builds up on the soil surface, following detachment, overland water flow transports suspended soil downslope. Suspended soil in the runoff water abrades and detaches additional soil particles as the water travels overland. Preventing detachment is the most effective point of erosion control because it keeps the soil in place. Other erosion control practices seek to slow soil particle transport and cause soil to be deposited before it reaches streams. These methods are less effective at protecting the quality of soil within the field.

Commonly implemented practices to slow soil transport include terraces and diversions. Terraces, diversions, and many other erosion "control" practices are largely unnecessary if the ground stays covered year-round. For erosion prevention, a high percentage of ground cover is a good indicator of success, while bare ground is an "early warning" indicator for a high risk of erosion. Muddy runoff water and gullies are "too-late" indicators. The soil has already eroded by the time it shows up as muddy water, and it's too late to save soil already suspended in the water.

Protecting the soil from erosion is *the* first step toward a sustainable agriculture. Since water erosion is initiated by raindrop impact on bare soil, any management practice that protects the soil from raindrop impact will decrease erosion and increase water entry into the soil. Mulches, cover crops, and crop residues serve this purpose well.

Additionally, well-aggregated soils resist crusting because water-stable aggregates are less likely to break apart when the raindrop hits them. Adequate organic matter with high soil biological activity leads to high soil aggregation.

Many studies have shown that cropping systems that maintain a soil-protecting plant canopy or residue cover have the least soil erosion. This is universally true. Long-term cropping studies begun in 1888 at the University of Missouri provide dramatic evidence of this. Gantser and colleagues examined the effects of a century of cropping on soil erosion. They compared depth of topsoil remaining after 100 years of cropping. As the table shows, the cropping system that maintained the highest amount of permanent ground cover had the greatest amount of topsoil left.

Table 7. Topsoil depth remaining after 100 years of different cropping practices.

Crop Sequence	*Inches of topsoil remaining*
Continuous Corn	7.7
6-year rotation	12.2
Continuous timothy grass	17.4

The researchers commented that subsoil had been mixed with topsoil in the continuous corn plots from plowing, making the real topsoil depth less than was apparent. In reality, all the topsoil was lost from the continuous corn plots in only 100 years. The rotation lost about half the topsoil over 100 years. How can we feed future generations with this type of farming practice?

In a study of many different soil types in each of the major climatic zones of the U.S., researchers showed dramatic differences in soil erosion when comparing row crops to perennial sods. Row crops consisted of cotton or corn, and sod crops were bluegrass or bermuda grass. On average, the row crops eroded more than 50 times more soil than did the perennial sod crops. The two primary influencing factors are ground cover and tillage.

So, how long do fields have before the topsoil is gone? This depends on where in the country the field is located. Some soils naturally have very thick topsoil, while other soils have thin topsoil over rock or gravel. Roughly 8 tons/acre/year of soil-erosion loss amounts to the thickness of a dime spread over an acre. Twenty dimes stack up to 1-inch high. So a landscape with an 8-ton erosion rate would lose an inch of topsoil about every 20 years. On a soil with a thick topsoil, this amount is barely detectable within a person's lifetime and may not be noticed. Soils with naturally thin topsoils or topsoils that have been previously eroded can be transformed from productive to degraded land within a generation.

Tillage for the production of annual crops is the major problem in agriculture, causing soil erosion and the loss of soil quality. Any agricultural practice that creates and maintains bare ground is inherently less sustainable than practices that keep the ground covered throughout the year. Wes Jackson has spent much of his career developing perennial grain crops and cropping systems that mimic the natural prairie. Perennial grain crops do not require tillage to establish year after year, and the ground is left covered. Ultimately, this is the future of grain production and truly represents a new vision for how we produce food. The greatest research need in agriculture today is breeding work to develop perennial crops that will replace annual crops requiring tillage. Farming practices using annual crops in ways that mimic perennial systems, such as no-till and cover crops, are our best alternative until perennial systems are developed.

REFERENCES

Finch, C.V. and C.W. Sharp. 1976 *Cover Crops in California Orchards and Vineyards.* USDA Soil Conservation Service, Washington, D.C.

Hendrix, P.H., D.A.Jr. Crossley and D.C. Coleman. 1990. "Soil biota as components of sustainable agroecosystems". In C.A. Edwards, Lal, Rattan, P. Madden, R. Miller, H. House, Gar (eds.) *Sustainable Agricultural Systems.* Soil and Water Conservation Society, IA.

Hudson, Berman E. 1994. "Soil organic matter and available water capacity". *Journal of Soil and Water Conservation.* Volume 49, Number 2. p. 189-194.

Magdoff, F.R. 1992. *Building soils for better crops: organic matter management.* University of Nebraska Press, NE.

Nair, P.K.R. 1982. *Soil Productivity Aspects of Agroforestry.* ICRAF, Nairobi.s

4

Organic Crop Production

Organic agriculture as farming without synthetic pesticides and conventional fertilisers. This should not be considered a definition but a characteristic—only one characteristic of a socially and environmentally conscious approach to agriculture that is currently experiencing rapid growth. Research on organic farms, done over several decades, has revealed characteristics usually associated with sustainable farming, such as reduced soil erosion, lower fossil fuel consumption, less leaching of nitrate, greater carbon sequestration and, of course, little to no pesticide use.

In 2002, when the United States Department of Agriculture (USDA) adopted the National Organic Standard that spells out what farmers and food processors must do to be certified "organic," the organic industry already had a long history of relying on third-party certifiers to ensure the integrity of their products and practices. Under this system, a state-run or accredited private agency evaluates farmers and processors to see whether they conform to the standards of the National Organic Program (NOP). Those who do can then market their products as "USDA Certified Organic" and display the official USDA organic seal on their packaging. In essence, certification is largely about integrity—assuring that the buyer is getting what he or she is paying for.

Certified organic production, then, means production by approved organic methods, with additional pains taken to eliminate contamination with prohibited materials and commingling with conventional products. There is a common misconception that certified organic means "pesticide residue-free."

Consumers have a right to expect little or no pesticide residue on certified organic crops because none are used in their production. However, ours is a dirty world in which pesticides and their break-down products are ubiquitous. This is only to be expected in a national farm system where more than 99% of all applied farm chemicals miss the target organism.

ORGANIC PRINCIPLES

There are several compelling principles that characterise certified organic farming. They include biodiversity, integration, sustainability, natural plant nutrition, natural pest management, and integrity. Most organic operations will reflect all of these to a greater or lesser degree. Since each farm is a distinct entity, there is a large degree of variation.

Biodiversity

As a general rule, diverse ecosystems in nature have a higher degree of stability than those with only a few species. The same is essentially true for agroecosystems. Farms with a diverse mix of crops have a better chance of supporting beneficial organisms that assist in pollination and pest management.

Diversity above ground also suggests diversity in the soil, providing better nutrient cycling, disease suppression, tilth, and nitrogen fixation. Good organic farmers mimic the biodiversity of nature through practices like intercropping, companion planting, establishment of beneficial habitats, and crop rotation. The effort to increase biodiversity works hand-in-hand with enterprise diversity, which is often an objective on organic farms.

Diversification and Integration of Enterprises

The drive to build biodiversity in organic systems encourages diversity among enterprises, but not as isolated or independent entities. Good organic operations integrate their various enterprises. A good example can be seen among Midwestern organic crop and livestock operations. The typical Midwestern organic operation ties the needs of crops and livestock together in a practical and elegant way. The forage and grain needs of ruminant livestock make for a diverse mix of crops. Particularly valuable is the inclusion of legume forages for ruminant feed. Forage legumes in rotation fix a sustainable supply of nitrogen in the soil that feeds subsequent non-legume crops in rotation. Manure from the livestock enterprises is conserved as a nutrient resource and recycled back to the crop fields. Farms such as these have the additional advantage of greater economic sustainability, as their risks are spread over several livestock and crop enterprises.

Sustainability

In addition to the greater economic sustainability afforded by enterprise diversification, organic farmers are often able to reap market premiums for certified production. However, since many organic enterprises realise somewhat lower marketable yields, this has not always translated into higher profits or greater economic sustainability. As more and more organic growers enter the marketplace, it is likely that premiums will stabilise at modest levels and may vanish for some crops.

Organic producers need to look well ahead and be aware of shifting trends. Many organic farms perform well on many of the measurable indicators associated with sustainability, such as energy consumption and environmental protection. However, sustainability is an ideal, and the best that can be said is that current organic farms are closer to the ideal than most alternatives—certainly closer than comparable conventional farming operations. The extent to which traditional organic agriculture philosophy influences the adoption of sustainable practices has only been touched upon.

During the Washington University studies of midwestern farms in the late 1970s, researchers observed that organic farmers had embraced conservation tillage technologies at a much faster rate than their conventional counterparts. Conservation tillage was not and is not considered a traditional practice of organic farming, yet its ready adoption points to the dynamic nature of organic agriculture and offers clear evidence that the underlying philosophy of sustainability—strongly championed by Albert Howard—remains a vital part of organics. Given the option of a sustainable technology that fits the constraints of certified organic agriculture, it is natural for most organic farmers to choose it.

Natural Plant Nutrition

Even though we require the same basic "stuff" to live, it is somewhat challenging to draw simple comparisons between the nutritional needs and processes of plants and those of animals. Plants are able to photosynthesize to make sugars, which are ultimately synthesized into proteins and other plant constituents. Humans and other animals, by contrast, can obtain energy foods, proteins, and vitamins only by consuming plants or other animals. Both plants and animals also require minerals. Humans and other animals extract minerals, along with sugars and proteins, from the food they eat.

Plants, too, obtain minerals—and a wide range of vitamins, antibiotics, and other useful compounds—through digestion. However, plant digestive systems are not internalised as they are in animals. Plants must rely on the external digestive processes of the soil system within reach of their roots—a zone called the rhizosphere.

Nutrient Absorption

Critics are often under the illusion that organic farmers believe plants obtain all their nutrients from an organically managed soil in a chemically organic form. While a few organicists may believe that, the majority recognises that digestion processes in the soil release minerals in forms similar to those applied as commercial fertilisers.

Unfortunately, the notion that organic farmers are naïve and ignorant about basic agronomy is a red herring that has often foiled intelligent discussion about the pros and cons of the system. Among the facts that are often obscured is that plants can and do absorb significant amounts of large organic molecules from the soil; herbicide and systemic insecticides are among these. In healthy soils they also absorb vitamins, chelated minerals, hormones, and other beneficial compounds. The organic philosophy of crop nutrition begins with proper care and nourishment of the organisms responsible for the soil digestive process.

Organic farmers believe this is best accomplished by avoiding toxic chemicals and practices—like excessive tillage—that are harmful to soil organisms, as well as by the addition of organic matter and natural rock minerals. Conventional systems, in contrast, try to circumvent the soil's digestive process and provide needed minerals to the plant directly, in a soluble form. From the organic perspective, the conventional approach has several flaws.

- Applying large quantities of soluble fertiliser to a crop only one, two, or three times per season floods the plant with those nutrients, causing nutritional imbalances that lead to crop diseases, insect infestations, and reduced food quality.
- Failure to support and care for soil biotic life, along with other practices that are downright destructive, ultimately leads to its decline. As a result, plants lose out on the vitamins and other beneficial products these organisms produce, tilth is reduced, and the soil becomes increasingly dependent on synthetic inputs.
- Conventional fertilisation tends to concentrate on a limited number of macronutrients, even though the need for at least 13 soil minerals is scientifically recognised. This skewed focus is also responsible for generating imbalances in the plant.
- Application of large amounts of soluble nutrients can stimulate certain problem weed species.
- Soluble nutrients—especially nitrate—are prone to leaching, which can cause a number of environmental and health problems.It is organic farming's approach to soil building and plant fertilisation that is the true basis for the belief that organic food and feed has superior nutritional value, much more so than the absence of pesticide residues, which has drawn the spotlight ever since the 1960s.

Natural Pest Management

Whether conventional or organic, all farmers are concerned with pests. They spend a lot of time and resources controlling them. However, in the organic "world view," pests—whether weeds, insects or diseases—are not simply scourges. They are indicators of how far a production system has strayed from the natural ecosystems it should imitate. Certain weeds, for example, tend to predominate when soils are too acidic or too basic; some become a problem when soil structure is poor and conditions become anaerobic; others may be stimulated by excessive fertiliser or manure salts.

Organic proponents also believe that insect pests are attracted to inferior or weak plants—the result of poor crop nutrition. Their logic continues by asserting that pests are naturally repelled by vigorous, well-nourished plants. This belief is often challenged, and significant research remains to be done. As scientific understanding has grown, insect pest outbreaks are also being understood as imbalances in the whole agroecosystem and how it is managed. In nature, massive pest outbreaks are relatively rare and short-lived, due to the presence of natural predators, parasites, and disease agents that quickly knock the pest numbers back down to a moderate level.

In farming systems that inadvertently destroy or otherwise fail to support the natural control complex, pest problems are routine and, typically, worsen with time. The farmer becomes increasingly addicted to costly and extreme control methods to produce a crop. Most organic growers consider pesticides to be a cause of agroecosystem imbalances and employ allowed natural pesticides as little as possible.

Integrity

Integrity refers to the systems in place and actions undertaken to assure that consumers of organic products get what they pay for. Consumers have a right to expect that the organic food they buy not only be raised by organic methods but be protected from contamination and from commingling with non-organic products.

While the responsibility for much of this rests with others in the organic marketing chain, many certified organic growers need to incorporate additional practices that work to assure the integrity of their products. Proper record keeping is very important in this regard, though growers are often reluctant to spend much time on it. Among the more important production practices in the field are buffer strips, which reduce chemical drift from neighboring fields and roadsides, while also serving water and soil conservation objectives.

Tools and Practices

The tools and practices of organic agriculture include traditional alternatives—crop rotation, manuring, liming, etc.—long recognised as important to a sound production system. They also include more contemporary practices and materials that research and keen observation over time have contributed. Note, too, that each farm operation will employ its own combination of tools and practices to build a working organic system. There is no simple cookbook formula for combining them in ideal proportions.

Planned Crop Rotation

Essentially a tool for annual cropping systems, crop rotation refers to the sequence of crops and cover crops grown on a specific field. Particular sequences confer particular benefits to long and short-term soil fertility, and to pest management.

Agronomic operations are especially dependent on crop rotations that include forage legumes. These provide the vast majority of the nitrogen required by subsequent crops like corn, which is a heavy consumer of that nutrient. Even when livestock are present to generate manure, the animals are largely recycling the nitrogen originally fixed by legumes in the system. The basic Midwestern rotation demonstrates the elegant way in which a whole farm system can function:

- Legumes fix nitrogen in the soil, providing for subsequent non-legumes in the rotation.
- Several insect pest cycles are interrupted, especially those of the northern and western rootworm species that can be devastating to corn.
- Several plant diseases are suppressed, in cluding soybean cyst nematode.
- Weed control is enhanced as perennial weeds are destroyed through cultivation of annual grains; most annual weeds are smothered or eliminated by mowing when alfalfa is in production.
- Livestock manures are applied just in advance of corn, a heavy nitrogen consumer.
- All crops can be marketed either as is or fed to livestock to be converted into value-added milk, meat or other livestock products.

Green Manures and Cover Crops

Green manuring consists of incorporating into the soil a crop grown for the purposes of soil improvement. It is a practice with a long history. Green manuring has been ignored in recent years as a serious option for soil improvement because the traditional practice entailed planting a full-season cover crop. This removed the field from commercial production for a whole season.

Interest has returned, however, since green manuring strategies have been combined with cover cropping schemes. Cover cropping is growing a crop for the purpose of soil and nutrient conservation. It is a more contemporary concept than green manuring, in crop agriculture. The two concepts—cover cropping and green manuring—go well together, as most cover crops are easily used as green manures prior to the planting of a commercial crop.

The combined benefits become economically feasible when the cover is grown during the off-season or inter-seeded with the main crop. It is made even more desirable when the cover crop includes nitrogen-fixing legumes.

Manuring and Composting

Livestock manures are the most traditional and widely recognised organic fertilisers. Under ideal circumstances, livestock enterprises are integrated into the whole farm operation, and manuring becomes part of a closed system of nutrient recycling. This is still strongly encouraged in organic operations. In reality, however, crops and livestock production are often divorced from each other, and manures must be imported.

This has created some concerns in the organic community, since much manure is now generated by large, industrial agriculture feeding operations called CAFOs (Confined Animal Feeding Operations). Not only are there concerns about contaminants but many in the organic community also object to any "partnering" with this segment of conventional agriculture, which is considered at odds with the environmental and social values represented by organic farming. Nonetheless, the National Organic Program does not differentiate between CAFO and other livestock manure sources. However, the NOP regulations do require that livestock manure not contain any synthetic substances not included on the National List of synthetic substances allowed for use in organic crop production. Another issue that has grown up around manure use in organic farming relates to food safety.

One of the best means of handling manures is composting. Composting stabilises the nutrients in manure, builds populations of beneficial organisms, and has a highly beneficial effect on soils and crops. Compost can be produced on-farm by a number of means. Additional products from composts, such as compost teas, have special applications in organic agriculture. Human manures are expressly forbidden in certified organic production. This includes composted sewage sludge. The organic community made its opinion on this quite clear when the USDA's first draft of the national rule proposed allowing the use of sludge in certified production. It was counted as one of the "big three" targets of protest, along with food irradiation and genetic engineering. The prohibition of biosolids would have been disconcerting to Albert Howard, who decried the failure of cities to return their organic wastes to the countryside. Such recycling was, in his mind, a key aspect of sustainability.

Livestock on Organic Farms

Among the thorniest of issues swirling around the edges of organic agriculture is the role of livestock. The disagreements arise because of the diversity of people and philosophies in the organic community. Organic agriculture can usually count vegetarians and animal welfare proponents among its more vocal supporters. Many of these people feel strongly that animals should not be exploited. Their rationale often goes beyond emotional and religious beliefs; convincing human health concerns, social issues, and environmental reasons are commonly cited.

On the other side of this argument are those who feel that an organic farm cannot achieve its full potential or ecological balance without livestock manure; that it is essential to nutrient cycling and to the finer aspects of soil building. Excellent soil fertility can be built in the absence of farm livestock and livestock manures by using vegetation-based composts and by harnessing the livestock in the soil—earthworms and other soil organisms. However, it is clearly easier to design a contemporary, low-input organic farm when traditional livestock are integrated. The biological and enterprise diversity that livestock can bring contributes enormously to stability and sustainability.

Many in both the organic and animal welfare communities are working to prohibit factory farming of livestock in organic systems. Many of the difficulties revolve around the fine interpretations of language in various organic standards. Wording such as "access to fresh air and sunlight," for example, can be construed to mean nothing more than opening the door on one end of a large confinement poultry house for a couple hours a day.

What Howard had not taken into account is the almost universal contamination of urban wastes with heavy metals and chemicals that are not eliminated by composting and may even be concentrated. Perhaps this was not yet a serious problem in his time; it is, however, in ours.

Organic farmers and consumers concerned about contamination of soil and crops with agricultural pesticides and synthetic fertilisers would be remiss to ignore the contamination hazards of even well-composted sewage.

Intercropping and Companion Planting

Interplanting two or more mutually beneficial crops in close proximity is one strategy for increasing biodiversity. In large-scale mechanised crop culture, this is called intercropping. It typically involves alternating rows or a number of rows of compatible field crops, like soybeans and corn. It also applies to sowing multiple forage crops, like alfalfa, bromegrass, and timothy, when these are grown together. When interplanting is done on a smaller scale, it is often called companion planting. A classic example of companion planting is the inter-planting of corn with pole beans and vining squash or pumpkins. In this system, the beans provide nitrogen; the corn provides support for the beans and a "screen" against squash vine borer; the vining squash provides a weed suppressive canopy and discourages corn-hungry raccoons.

Biological Pest Control

Organic farming relies heavily on populations of beneficial insect predators and parasites, pest disease agents, insect-eating birds and bats, and other creatures, to help manage pest problems. These biological controls help keep pest numbers at levels where further cultural activities or relatively mild pesticides are adequate to assure a crop. In some instances, biological control can be so effective that no additional action is even needed by the farmer. Some see biological control as a default benefit of the soil fertility practices of organic farming.

The diversity of crops in a soil-building rotation, the use of cover crops, and other practices build a diverse soil biology that works to keep soil pests in check. They also provide substantial above-ground habitat for beneficials. The absence of pesticides also favors biocontrol. In many organic systems, farmers sometimes purchase and release control agents like ladybird beetles, lacewings, trichogramma wasps, etc., or use weeder geese—a quaint but effective biological weed control. Increasingly, growers are designing and maintaining both permanent and temporary habitats specifically for beneficial insects, spiders, and other helpful species. This is known as *farmscaping*.

Sanitation

Sanitation can take on many forms:

- removal, burning, or deep plowing of crop residues that could carry plant disease or insect pest agents
- destruction of nearby weedy habitats that shelter pests
- cleaning accumulated weed seeds from farm equipment before entering a new, "clean" field
- sterilising pruning tools

As in human and animal health, sanitation practices can go a long way in preventing crop pest problems. However, many practices—such as clean cultivation, deep plowing, and burning crop residues—can increase erosion and reduce biodiversity. Thus, they may conflict with sustainability. Good organic growers recognise this and treat those practices as transitional or rescue options, rather than relying on them on an annual basis.

Tillage and Cultivation

Tillage and cultivation are tools that can accomplish a variety of objectives in farming systems: weed control, crop residue management, soil aeration, conservation of manures and other fertilisers, hardpan reduction, sanitation to destroy pest and disease habitat, etc. While conventional farmers rely on chemicals to accomplish many of these objectives, organic growers focus more on improving tillage and maximising its benefits.

Guidelines for primary tillage, for example, are intent on conserving crop residues and added manures in the upper, biologically active zones of the soil, rather than burying them deeply where decomposition is anaerobic. Leaving soils completely bare and vulnerable to erosion is discouraged; fall moldboard plowing is certainly frowned upon. Cultivation in organic systems often rises to the level of art. Row-crop farmers frequently use blind cultivation -shallow tillage, which largely ignores the crop rows-beginning shortly after seeding until the plants are but a few inches high.

Rotary hoes, wire-tooth harrows, and similar equipment can be used for blind cultivation, delaying the first flush of weeds and giving the crop a head start.

Conservation Tillage

Organic agriculture is often characterised as addicted to maximum tillage—with growers using every opportunity to lay the land bare with shovel, plow, or rototiller. This image has been magnified through the popularity of small-scale organic systems like the French Intensive and Biointensive Mini Farming models that espouse double- and triple-digging to create deep rooting beds.

While appropriate to such intensive systems, this degree of cultivation is not characteristic of organic agriculture in general. It may surprise some to learn that a large number of organic producers are not only interested in conservation tillage, they have adopted it. This will be a surprise because many believe that conservation tillage always requires herbicides. The interest in conservation tillage among organic producers in the Cornbelt was well documented in the mid-1970s by Washington University researchers. They noted that the vast majority of organic farmers participating in their studies had abandoned the moldboard plow for chisel plows.

Plowing with a chisel implement is a form of mulch tillage, in which residues are mixed in the upper layers of the soil, and a significant percentage remains on the soil surface to reduce erosion. Furthermore, a notable number of organic farmers had gone further to adopt ridge-tillage, a system with even greater potential to reduce erosion. It was especially interesting to note that the use of these conservation technologies was almost nil among neighboring conventional farms at the time.

After blind cultivation, subsequent weed control operations in larger-scale systems can make use of advances in tillage equipment such as rolling cultivators, finger weeders, and torsion weeders that allow tilling close to the plant row. Smaller-scale operations often use wheel hoes, stirrup hoes, and other less capital-intensive hardware.

Determining the amount, the timing, and the kind of tillage to be done can be a balancing act for the organic grower, but experience and observation over time lead to proficiency. There are downsides to tillage, however, and most organic growers are well aware of them. The most obvious of these is the dollar cost; organic farmers are as concerned as their conventional counterparts about costs of production and strive to minimise expensive field operations. There is also a cost to the soil and environment.Every tillage operation aerates the soil and speeds the decomposition of the organic fraction. While this may provide a boost to the current crop, it can be overdone and "burn up" the humus reserves in the soil. Excessive tillage can also be directly destructive to earthworms and their tunneling, reducing their benefits to the land. There is also the danger of compaction, even when field operations are well timed.

Mulching

Mulching is a practice often used by organic growers. Traditionally, it entails the spreading of large amounts of organic materials—straw, old hay, wood chips, etc.—over otherwise bare soil between and among crop plants. Organic mulches regulate soil moisture and temperature, suppress weeds, and provide organic matter to the soil. Mulching is most appropriate to small, intensive operations with high-value annual or fruit crops.

A few systems of no-till organic gardening have evolved from the concept of deep, permanent mulching. Among these are the well-known Ruth Stout method and Synergistic Agriculture—a raised bed system developed by Emilia Hazelip, who adapted concepts from Permaculture and the ideas of Masanobu Fukuoka. Mark Cain and Michael Crane, co-owners of Dripping Springs Gardens—an intensive market gardening operation in Arkansas—have adapted Emilia's system to their farm with considerable success.

Plastic mulch, as long as it is removed at end of growing or harvest season, is also permitted in certified organic production. Its use allows larger acreage to be brought more easily under herbicide-free management, though there are serious issues to be addressed.

High-Input Organic Agriculture

Organic farming was described as a system that uses a minimum of off-farm inputs. While that describes most of organic agriculture as it is currently practiced, certified organic farming can also entail much greater reliance on off-farm inputs. Intensive annual strawberry and vegetable systems under plasticulture are good examples. In these systems, traditional rotations and soil building practices are usually employed, followed by clean cultivation and the laying of plastic mulch and drip irrigation tape on shaped beds.

During the season, large amounts of soluble organic fertilisers—typically fish-based—are fed to the crop through the drip system. At the end of the season, all plastics must be removed from the field, and it is returned to more standard organic management. Ideally, an off-season cover crop will be planted. Such systems are often exceptionally productive and economically attractive, when organic premiums are good. The high cost of soluble organic fertiliser, however, plus the marginally higher cost of pest controls, make such systems largely non-competitive in the conventional marketplace.

The labeling of such high-input systems as organic presents a paradox for many proponents of organic agriculture. It is unclear whether these technological advancements reflect the kind of farming most practitioners and supporters of organics think of as truly "organic." To begin with, the research citing environmental and economic benefits has largely been done on low-input organic systems; it is questionable whether similar findings would be made about high-input systems, especially regarding environmental matters. Of particular note, while low-input organic systems are documented as being more resistant to erosion, fields under plastic mulch are reported to be fifteen times more erodible.

Traditional organic farms leach minimal amounts of nitrogen into tile or groundwater; the losses from fields loaded with high levels of soluble organic fertilisers is certain to be greater, but how much greater is unknown. The fossil fuel energy involved in plastic manufacture, transportation, and application may or may not be compensated by reductions in tractor fuel use. Finally, the lowered capital investment required to produce a crop by traditional organic methods makes this form of farming more accessible to resource-poor farmers and entails less risk in years of crop failure or lack of premiums. These factors are less certain in a high-input system. A further consideration is the issue of plastic disposal following removal. At this time, there are few to no options for recycling, and landfills are the fate of plastic mulches at the end of each season.

While it is unwise to rush to judgement regarding high-input organic farming, it is clear that some adaptations will need to be made, if the traditional character and sustainability benefits of organic farming are to be preserved.

Fire

While fire can be used in a number of ways in organic agriculture, the area of greatest interest is flame or thermal weeding. In its most common application, torches mounted on a tractor toolbar throw a hot flame at the base of mature plants, over the inter-row area, or both. Tractor speed is adjusted so that weeds are not burned so much as seared. Searing is sufficient to kill most seedling weeds and uses less fuel. Liquid propane (LP) gas is the fuel most commonly used, though alternatives such as alcohol and methane offer the possibility of on-farm sources.

Supplemental Fertilisation

In many organic systems, crop rotation, manuring, green manuring, along with enhanced biological activity in the soil, provide an abundant supply of plant-essential minerals annually. This is especially true on naturally deep and rich prairie soils. It is less true on poorer soils and on those that have been heavily exploited through non-sustainable farming practices. To correct mineral deficiencies in organically managed soils, organic growers often apply ground or powdered rock minerals.

The most commonly used rock mineral is high-calcium aglime. Dolomitic limestone, various rock phosphates, gypsum, sulfate of potash-magnesia, and mined potassium sulfate are also common. These are all significant sources of primary (P, K) and/or secondary (Ca, Mg, S) plant nutrients. The savvy organic grower applies significant amounts of these materials only with the guidance of regular soil testing.

Less common are other rock powders and fines that are limited sources for the major nutrients but are rich in micronutrients or have some other soil-improving characteristic. Among these are glauconite (greensand), glacial gravel dust, lava sand, Azomite®, granite meal, and others.

Supplementary nutrients that include nitrogen are often provided in the form of animal or plant products and by-products such as fish emulsion, blood meal, feather meal, bone meal, alfalfa meal, and soybean meal. Most of these products also supply some organic matter, though that is not the primary reason they are used.

ORGANIC FARMING: TOOLS AND PRACTICES

One basis on which to evaluate the tools or practices one chooses for an organic operation is whether or not they contribute to biodiversity—biodiversity being one of the principal characteristics of a sustainable organic agriculture. Crop rotation, cover cropping, farmscaping, companion planting, and intercropping are outstanding examples of practices that contribute to biodiversity. They therefore contribute to the long-term stability and sustainability of the farm agroecosystem.

Composting and manuring likewise contribute to biodiversity, but since the diversity they promote is mostly in soil biota, it is rather less obvious to the casual observer. On the other end of the spectrum are practices such as tillage, cultivation, thermal weeding, solarisation, and plastic mulching. Such tools significantly reduce diversity in the field and tend to move the system in a less sustainable direction. This, however, does not necessarily make these practices bad choices.

Organic farming has often been called natural farming, as it tries to mimic the processes of nature in producing crops and livestock. However, the analogy goes only so far, since most agricultural systems are characterised by a struggle between human and nature, each with a clear notion of what plants and animals the land ought to support and in what proportions.

The farmer is typically in the position of "holding back" the natural succession of plant and animal species through the use of diversity-reducing tools and practices. The ideal is to bring about an agricultural system in which the long-term direction emphasizes diversity and sustainability. Among the best visualisations of this ideal are those emerging from the Permaculture movement.

Balanced Nutrition

Whether or not it has been customary in the past, organic growers are encouraged to have periodic soil testing done on their fields. How the results of a soil test are used, however, can vary considerably among farmers, depending on their personal philosophy and management skill. While there is certainly a segment of the organic farming community that has no faith in soil audits, most growers use them as a means to monitor progress in building their soils, to identify nutrient deficits, and to guide supplementary fertilisation. While there are many ideas about fertilisation guidelines, there are two schools of thought that dominate.

The first is commonly known as the sufficiency approach or model. Though somewhat oversimplified, the following are among its significant characteristics:

— annual fertiliser additions of P and K are based in good part on how much the crop is expected to remove from the soil at harvest

— additional amounts of P and K are recommended based on keeping the soil nutrient reserves at a particular level

— lime is added to the soil based on pH

— little to no attention is paid to nutrient balance or to the levels of secondary nutrients Ca and Mg

The second approach is referred to variously as cation nutrient balancing, the Albrecht system, the CEC, or the base saturation approach. It differs from the sufficiency approach in that fertiliser and lime recommendations are made based on an idealised ratio of nutrients in the soil and its capacity to hold those nutrients against leaching.

Cation nutrient balancing is more popular among practitioners of organic farming and sustainable agriculture in general than it is among conventional growers. However, there is no universal agreement on which approach is most appropriate for organic management. A wide range of other products -humates, humic acids, enzymes, catalyst waters, bioactivators, surfactants, to name a few—are also acceptable in organic crop production. However, the tradeoff between out-of-pocket cost and efficacy of such materials is often challenged by conventional and organic growers alike. Organic growers are encouraged to experiment, but to do so in a manner that allows the actual results to be measured.

Biorational Pesticides

While, in principle, any pesticide use is discouraged in organic systems, a rather wide range of biorational pesticides is permitted. The frequency of pesticide use varies considerably with crop and location. For example, there is virtually no use of pesticides on organic row crops in the Cornbelt. By contrast, organic tree fruits in the Midsouth routinely receive heavy applications of several fungicides and insecticides allowable in organic production. The pesticides permitted in organic farming fall predominantly into several classes.

Minerals: These include sulfur, copper, diatomaceous earth, and clay-based materials like Surround®.

Botanicals: Botanicals include common commercial materials such as rotenone, neem, and pyrethrum. Less common botanicals include quassia, equisetum, and ryania. Tobacco products like Black-Leaf 40® and strichnine are also botanicals but are prohibited in organic production due to their high toxicity.

Soaps: A number of commercial soap-based products are effective as insecticides, herbicides, fungicides, and algicides. Detergent-based products are not allowed for crop use in organic production.

Pheromones: Pheromones can be used as a means to confuse and disrupt pests during their mating cycles, or to draw them into traps.

Biologicals: One of the fastest-growing areas in pesticide development, biopesticides present some of the greatest hope for organic control of highly destructive pests. Among the most well-known biopesticides are the Bacillus thuringiensis (Bt) formulations for control of lepidopterous pests and Colorado potato beetle.

One of the greater difficulties that organic producers face on a regular basis is determining whether or not a particular product or material can be used in organic production. Sad to say, the problems are real, but some basic clarifications will help. First of all, all natural or nonsynthetic materials can be assumed to be acceprable in organic production. There are a few exceptions, however, which will be explained shortly.

An important organisation to know about is the Organic Materials Review Institute (OMRI). OMRI is a non-profit organisation that evaluates products for suitability in organic production and processing. OMRI does not have status as a regulatory body. However, its decisions with regard to the acceptability of commercial products are highly respected and accepted by most certifiers.

Of Seeds, Seedlings, and GMOs

One of the challenges faced by organic growers is getting suitable planting stock. Certified production requires that seed not be treated with pesticides and be organically grown. Transplants must be purchased from a certified organic source or otherwise be grown using organic methods. This presents some difficulty, as most commercial potting mixes contain prohibited fertilisers and wetting agents, requiring special ordering or the added effort of making a homemade mix.

Generally, planting stock for perennial crops like tree fruits and berries is available from conventional sources, even if treated with pesticides. However, no production can be sold as organic for a minimum of 12 months following transplanting to an organic field. In all instances, organic growers are not permitted to use varieties of crops that have been developed through genetic engineering (GE). At first glance, this might seem perplexing.

GE crops promise further means of non-chemical pest control and the possibility of nutritionally enhanced foods. However, the organic community is concerned about environmental, economic, and social impacts from this new technology which, they feel, have not been adequately studied. This same concern applies to GE foods. Natural foods consumers—a large segment of the organic market—do not want them. This was made abundantly clear to the USDA when its first draft of the proposed rule, which suggested permitting genetically modified organisms (GMOs), was released in 1997. The backlash was so strong that not only GE crop varieties, but all inputs, such as GMO-derived biopesticides, are prohibited from organic production.

Unfortunately the GE issue does not resolve itself so simply for organic farmers. The proliferation of GMOs in the marketplace and across the landscape has created a host of new challenges that have particular implications for organic agriculture:

— *Genetic drift onto crops.* Pollen drifting from adjacent fields can contaminate organic crops. Some clients in the lucrative European Union market accept 0% GMO contamination; others accept only very low levels. Pollen from cross-pollinating crops like corn easily travels hundreds of yards and may be carried for miles under the right conditions. No reasonable amount of non-crop buffer can prevent contamination except in highly isolated locations.

– *Genetic drift onto soils.* Concerns have been expressed over the persistence of GE pollen in soils. Should contaminated soils be decertified?

– *Non-labeled products.* Should soybean, cottonseed, and corn gluten meals be prohibited as fertilisers, since these may come from GE varieties? Should pesticidal use of vegetable oils be prohibited?

– *Pest control problems.* The rapid proliferation of corn, cotton, and other crops featuring the Bt gene from Bacillus thuringiensis is likely to accelerate the development of resistance to this natural pesticide. Organic growers may lose one of their most useful pest management tools.

– *Domination of the food system.* In the late 1990s, the public was made aware of two specific developments of agricultural biotechnology dubbed the terminator and traitor technologies. Terminator technology is the genetic altering of seed so that the subsequent generation will not germinate. This prevents farmers from saving a portion of the harvested crop to replant the next year—a traditional practice among farmers worldwide. Traitor or suicide technology is the genetic altering of seed so that it will fail to germinate or mature unless a proprietary chemical is applied. Such technologies are authoritarian and advance the centralisation and industrialisation of the food system. Organic farmers perceive additional challenges from genetic drift impacts and difficulty in finding non-GE seed in the marketplace.

Foliar Fertilisation

Foliar fertilisation or feeding entails the application—via spraying—of nutrients to plant leaves and stems and their absorption at those sites. It is not specifically an organic practice, though it is commonly used by many organic growers. The fertiliser materials used are typically soluble fish- and seaweed-based products, naturally chelated nutrients, humic acid extracts, and teas made from plants, dried blood, manure, guano, or compost.

At first glance, the use of foliar feeding appears contradictory to the organic notion that one feeds the soil to feed the plant. Organic growers rationalise the use of this approach on two points.

1. Foliar feeding is strictly supplemental fertilisation; it is not used as a substitute for traditional soil building practices.
2. Foliar fertilisation is understood to increase the production of root exudates, which stimulates biological activity in the rhizosphere (soil area adjacent to plant roots). The soil bio-life gets considerable benefit in this indirect way from foliar feeding.

Esoteric Practices

There are a number of farming and gardening practices based on belief in a non-physical world closely aligned with our physical reality. Those who use these practices believe that conditions in this unseen realm influence, or even dictate, what happens on a material level. They take pains to understand these influences and adjust their farm activities.

In some instances, they act to influence events. Obviously, these practices are rooted more deeply in metaphysics than in the conventional sciences. The most well-known of these esoteric practices is the scheduling of planting or other field operations according to lunar or other astrological signs. Dowsing and other forms of divining may also be used to guide scheduling, fertiliser selection, and other facets of farming.

The use of potentised preparations, as done in Biodynamic™ farming, is also a metaphysical practice. Radionics, energy balancing towers, "medicine" wheels, prayer—these are further examples of practices used by some farmers to bring a tangible spiritual element into their farm operation. Esoteric practices like these are not inherently part of organic agriculture. In all likelihood, only a small-to-modest minority practice them or believe in their efficacy. Still, esoteric practices are sometimes associated with organic agriculture, and it is almost a given that practitioners of "energetic agriculture" are either certified organic or actively pursue a sustainable form of agriculture.

The rationale for this association of organic with the esoteric is not difficult to understand. Most metaphysical practices are founded on a world-view similar to that of ecology, in which everything is related to everything else. The difference is that ecology considers this interrelationship only on a hard physical level; metaphysics perceives it also on a non-physical level. This belief tends to cast humankind as part of nature, not merely as an inhabitant. This perspective would tend to make one more concerned about the environment and, therefore, more readily drawn to organics as a philosophy and pragmatic farming approach. To put it in a nutshell, it is natural for the metaphysically-inclined to choose organics; it is far less likely, however, that the average organic farmer is drawn to metaphysical practices. A further factor plays into the association between esoteric farming practices and organics. It relates to the fact that organic agriculture has, for many years, been marginalised and treated as pseudoscience by the mainstream—an experience shared by practitioners of metaphysical arts. It is not surprising for underdogs to seek common ground.

Buffers and Barriers

Field buffers are strips of sod or other permanent vegetation at the edge of or surrounding crop fields. There is considerable interest in both conventional and alternative agriculture in these buffers, since they assist in reducing soil erosion and improving water quality. If managed properly, some buffers also serve as beneficial insect habitat. In certified organic production, however, field buffers have an added purpose in reducing crop contamination from chemicals used on adjacent land.

Most agencies require a minimum 25-foot buffer along "uncontrolled" borders where there is a hazard of chemical use. Wider borders may be required where hazards are great—for instance, where adjacent farms have synthetic pesticides applied by aircraft.

ORGANIC FARMING: NICHE MARKET

The pioneers of organic farming considered organics the preferred direction for the whole of agriculture to take. It is likely that most contemporary proponents still hold that view. While recent growth in the organic industry is definitely encouraging, much of the impetus is tied to its growth as a niche market, not as a serious shift in the direction of mainstream agriculture. Unfortunately, a likely reason for the newfound "tolerance" of organic agriculture in many land grant universities and other formerly hostile environs is its perception as a niche market opportunity only. As such, it is not a serious threat to the status quo.

Is it practical and responsible to promote organic agriculture as the dominant approach to farming in the future? While some are quick to answer, "Yes!", most are buffaloed by the nagging question, "Can organic agriculture feed the world?" Questions about the productivity and the prospects of widespread starvation have long been an effective tactic for stonewalling of organics. While it is not the purpose and settle such a compelling issue, there are a few points that can and should be made regarding the "starving billions" scenario.

— Critics warn that, were organic farming to be adopted on a wider scale, per-acre agricultural productivity would decline sharply; that meeting food needs would necessitate the plowing of even more erodible hillsides and the draining of more wetlands. Such scenarios are often based on pre-chemical era yield data that ignore the advancements of modern organic farming. Contemporary research on organic systems—as limited as it is—indicates that the per-acre productivity of organic and conventional systems is not vastly different.

  Other scenarios are sometimes based on the perceived need for vast quantities of grain—particularly corn. Since overproduction of grain has led to prices below costs of production for several years, this is a questionable basis for argument. It also ignores the fact that most of the grain produced is fed to livestock and that many livestock species can eat forage instead. Ruminant livestock—cattle and sheep especially—are designed by nature to thrive on forage and can even be finished for market on pasture.

- The paucity of good research on the productivity of organic systems also spotlights the lack of practical research for organic producers. That organic farms are as productive as they appear to be is remarkable in light of minimal research and extension support for many decades. It has often been argued that, had comparable resources been put into organic research, its widespread feasibility would be unquestionable. This assertion is well-supported by the development, in recent years, of a number of alternative, organically acceptable pesticides, that are certain to expand organic production of many crops in regions where it had previously been near impossible on a commercial basis.

- Increasing agricultural production alone does not alleviate hunger. The amount of grain produced in the world in 1999 could, by itself, sustain 8 billion people—2 billion more than our current population. Total food production is estimated to provide each human being with at least 3500 calories per day.

  The issue of hunger, it appears, is not so much a lack of food, but lack of entitlement to food. People are shocked to learn that while many Bangladeshees starved during the 1974 floods, roughly 4 million tons of rice produced in that country were stacked in warehouses for want of buyers. The people were simply too poor to buy it. They are surprised to learn that Ireland exported grain during the Irish potato famine while 1 million of its citizens died and even more emigrated; that India regularly exports food and animal feed despite an estimated 200 million in starvation. Those lacking the ready cash to buy food or the resources to produce it themselves seem destined for hunger no matter what miracles agricultural technology provides.

  The world's nations will need to deal with issues of equity and democracy first, if hunger is ever to be effectively addressed.

- No agriculture can continue to feed a growing population if it depletes or fouls its resource base. The path undertaken by conventional agriculture is ultimately a dead end in this regard, though there is an almost mystical faith that genetic engineering and other complex technologies will always triumph. Agriculture needs to be sustainable. Therefore, those who promote organic agriculture as a true alternative are well advised to do their part in ensuring that certification and regulation does not create a "compliance agriculture" in which sustainability becomes little more than an afterthought.

Record Keeping

Few would consider record keeping as a tool for organic crop production. However, the documentation of how and where a crop was raised, what products were applied and when, which bin or cooler it was stored in, is of critical importance to establishing the integrity of the product. If the farmer cannot provide reasonable documentation that his or her crop was organically grown, that it has not been contaminated with chemicals, and that it has not commingled with a similar conventional product, then certification may well be denied.

As stated at the outset, the tools and practices described in detail here are by no means an exhaustive listing of organic options. Among those not mention but requiring mention are timing of planting and field operations to avoid pests or disrupt their life cycles, various forms of pest traps, physical barriers to pests, and increasing plant populations to enhance crop-weed competition. While there have been varying notions of organic farming over the years, the growth of the industry and the introduction of standards and certification have led to a clearer definition in recent years. That definition describes organics as a viable agriculture, based on sound farming practices, that does not include synthetic chemicals.

Certified organic agriculture can be further characterised by a set of principles that include biodiversity, integration, sustainability, natural plant nutrition, natural pest management, and integrity. These principles are expressed through the implementation of both traditional and cutting-edge farm practices. As the organic industry continues to grow and evolve, it faces many challenges, including the consequences of its own success.

Economic opportunities invite new players into the marketplace who may have little interest in sustainability or the positive social benefits many have come to associate with organics. Organic was synonymous with family farms and we are finding out that is changing. In fact, the organic is going to continue to grow. That doesn't mean that it is going to support family farms the way it has in the past. With the whole organic movement, we assumed that the social would go along with the environmental movement, and what we are finding out is no, that is not necessarily true, and even what they do environmentally is questionable. Whether certified organic farming will survive its own success and continue as a socially and environmentally responsible alternative, or merely become a parallel production system based on minimal compliance to standards, remains to be seen.

RESOURCE USE EFFICIENCY IN ORGANIC AGRICULTURE

Agronomic Performance

Comparisons of the performance of organic and conventional agriculture systems are meaningful only when made over an intergenerational period of time in order to assess the continued capacity of natural resources to sustain agriculture. High yields in non-organic systems are often exploitative systems that degrade land, water, biodiversity and ecological services on which food production depends. Most comparisons of the efficiency of alternative production systems focus merely on the gross yield of marketable commodities. Farmers usually experience a decline in yields after discarding synthetic inputs and converting their operations to organic production. After the agro-ecosystem is restored and organic management systems are fully implemented, yields increase significantly.

If conversion to organic departs from low-input (often traditional) systems, then yields tend to be stable. A study from Kenya indicates that organic agriculture in the tropics shows a good performance. This finding was contrary to conventional wisdom that contended such a system would be constrained by insufficient organic material. Organic systems in medium potential areas significantly out-performed conventional methods in maise grain yields, net cash benefits, return on capital and return per family labour day. For example, organically grown maize experienced less damage from weevils during storage than its conventional counterpart. This and other similar findings show that organic systems can double or triple the productivity of traditional systems.

The productivity of organic agriculture systems varies through the different stages of management:

(i) in-transition from conventional to organic management;

(ii) in-conversion from traditional to organic management;

(iii) organic management based on input substitution, and

(iv) complete shift to a systems approach.

The need to secure farm economic viability in the short-term results in few farms achieving a systems approach.

In most cases, transitioning farmers suffer decreased productivity, but even then there are cases where the conventional system is in such decline that there are significant positive yield responses to the addition of organic matter, particularly when there is a carryover of synthetic N fertiliser such as calcium nitrate. Also, there are cases where transitioning farmers still have some carryover with herbicides that suppresses weeds well into the second year of transition; for them the decline in yields related to weed pressure does not begin until the second or third year. This is most common in grain field crops such as wheat, rice and corn.

Yield losses are greatly exaggerated by conventional experiment stations. The actual experience of most transitioning farmers is that the yield losses are tolerable. The increased labour seems to hurt US farmers more than lower yields. Yield losses are caused by a number of interrelated factors: soil organic matter and biological activity take time to get established; many conventional farms are on a pesticide 'treadmill' that does not permit the establishment of beneficial organisms for pest, weed, and disease suppression; and fertility problems are common until restoration of full biological activity.

The degree of yield loss varies, and depends on inherent biological attributes of the farm, farmer expertise, the extent to which synthetic inputs were used under previous management, and the state of natural resources. Sometimes a few years may suffice or it may take many years to restore the ecosystem to the point where organic production is viable in economic terms. There are some soils that are so marginal and highly depleted that it would not be possible to farm without substantial government subsidies for certain commodities.

In the medium term, and as expertise increases, the value of organic agriculture becomes more evident because of yield improvements and increased fertility of the agricultural system. In the longer term, the performance of organic agriculture increases in parallel with improvements in ecosystem functions and management skills. The performance of organic agriculture cannot be judged based on the comparison of a single-crop or a single year. Organic agriculture usually performs better if one considers total production of useful crops per area. The greatest constraints faced by transitioning farmers are the lack knowledge, information sources, and technical support. Greater government investment in appropriate research and extension services can help overcome these constraints.

Extreme weather fluctuations present a growing threat to agriculture. Organic systems appear to be more stable and resilient in response to climate disruption based on comparisons with their conventional counterparts under stress conditions such as severe drought and flooding. In 1993, conventional rice in Japan was nearly wiped out by an unusually cold summer while organic farmers yielded 60-80 percent of the annual average. The better composition of water-stable aggregates in organic soils and reduced soil compaction result in the favourable performance of organic systems under both flood and drought conditions.

Yield comparisons offer a limited, narrow, and often misleading picture of the different production systems. Profitability and long-term economic viability would be a better indicator to determine what techniques operators would choose. Organic production, processing and marketing depend on several interrelated factors. Profitability in organic agriculture depends on relative input and labour costs, actual production costs, conventional and organic market conditions and the price premiums received for organic commodities. In some countries, government payments for the adoption of organic techniques are available to offset certain costs and encourage-practices that have long-term benefits to the environment. Accurate comparisons of economic performance must take account of performance over a complete rotation rather than a single year. Moreover, the multiple environmental benefits of organic farming, difficult to quantify in monetary terms, are essential ingredients in any comparison.

Economic performance

Only a few studies have assessed the long-term profitability of organic agriculture systems. While these studies vary in their methodologies and conclusions, they consistently show high revenues relative to conventional agriculture because of the premiums received. However, the costs relate to whole farm production over the whole rotation period: this includes both marketed products and non-food products. In particular, incomes achieved in one season may appear high because of price premiums but low in the subsequent rotation seasons if these crops have low or no market values. Looking at these seasons individually does not give an accurate picture of the financial viability of organic agriculture. Unfortunately, many comparative studies of organic and conventional or integrated production focus on a single crop and a single year.

The economic performance of organic agriculture in Europe shows a situation where organic farmers receive financial support and premium prices but where labour is expensive. An extensive analysis of European farm economics in terms of labour use, yields, prices, costs and support payments concluded that profits on organic farms are, on average, comparable to those on conventional farms. In the United States, profits are also comparable, despite the fact that there are no direct subsidies for organic agriculture.

In developing countries where organic agriculture is not subsidised, synthetic inputs are expensive and labour is relatively cheap, market-oriented organic farmers can achieve higher returns thanks to reduced production costs and diversified production. For example, in the Philippines, price premiums are not a sufficient incentive to market rice as organic. Farmers have adopted organic practices nevertheless because the avoidance of external inputs saves on production costs while yields are more stable. In Madagascar, hundreds of farmers have increased their irrigated rice yields from two to as much as eight tonnes per hectare by using local seeds, composts and innovative soil, plant, water and nutrient management practices.

Organic products tend to command impressive premiums in developed countries at retail level: on average between 10-50 percent (and in some cases as high as 100-200 percent) above conventional prices for the same commodity. These premiums reflect several underlying factors on both the demand and supply side. Premiums reflect strong consumer demand, with some consumers willing to pay higher premiums than others do. Most consumers in developed countries will pay a premium for organic, but only to a point. As the premium increases, the number of consumers willing to pay it decreases, because the conventional commodity is always available as a substitute.

Premiums in excess of 50 percent usually have underlying supply constraints and bottlenecks. These are often temporary and unpredictable. Because fewer fungicides and post-harvest tools are available, fresh organic produce tends to be more seasonal and local. Long-term high premiums often reflect severe production problems related to chronic endemic pests and diseases that cannot be managed effectively by existing biological and cultural techniques. On the other hand, situations where organic production costs are as low as, or lower than, conventional production will frequently see little or no organic price premium received by farmers.

Premiums compensate farmers for skilled resource management; higher labour costs, unit costs, and handling expenses; and administrative, inspection, and certification fees. Premiums also reflect the price to avoid and mitigate negative environmental externalities incurred by conventional agriculture. These include costs of damage to natural capital, human health and reduction of water, air and soil pollution. Such indirect costs are not usually included in food prices, and this distorts the market while encouraging activities that are costly to society.

Many reasons contribute to the additional costs to market organic products: inspection and certification fees, segregated storage, fewer options to control post-harvest pests and diseases, properly cleaned and well-documented transportation, careful handling to avoid dilution and contamination, appropriate packaging, and economies of scale. Because organic producers comprise a smaller proportion of the agricultural industry, individual producers are widely dispersed. Greater collection and assembly costs further add to the costs of transportation. Pest and disease infestations can result in handlers facing an unpleasant choice between losing most and possibly all of a crop, or treating it with a prohibited substance to recover the losses, and selling the product as non-organic. Segregation increases costs of handling. Many retailers require their suppliers to provide individual packaging and special labelling for organic food not required for conventional food.

To date, consumers in industrialised countries have been willing to pay a premium for organic food because they perceive environmental, health, or other benefits from that choice. While surveys show that consumer demand is unmet, organic farmers also report insufficient demand for their products. Balanced expansion of supply and demand and reduction in organic production costs (achieved through targeted research) will be one factor in maintaining the organic price premium. At present, the general tendency for demand to out-grow supply suggests that the premium is not under immediate threat for most product categories. Organic production is expected to continue to offer premium prices and a profitable alternative to conventional production systems for many farmers.

High prices and limited outlets have historically curtailed demand for organic agriculture. Lower prices would expand the organic market without discouraging producers, provided that the premium still compensates the costs of transition and provides a living wage to the producer. Most of the premium is captured by retailers, wholesalers, distributors and processors. Retailers can reduce prices while maintaining the profitability of organic farmers. A price premium to producers of 10-20 percent, perhaps even 50 percent, would have almost no impact on consumers. This, however, is not likely to be accepted by retailers and where feasible, direct marketing channels are being developed. At present, the marketing strategy of many major food retail chains is to expand the supply of a few low-cost organic products produced by a relatively small number of producers. This strategy benefits a few large organic farms, which rely on input substitution and global market over small and medium-sized local farmers.

Greater investments in research and extension offer long-term solutions to organic production and handling constraints. The redirection of only a small portion of public expenditures towards biological and cultural methods could increase yields, lower handling, distribution, and marketing costs, and deliver lower prices for organic food to consumers. As the organic market grows and matures, economies of scale should narrow margins for conventional products.

In many developing countries, there are no domestic institutions that can assist farmers to produce, handle, and market organic food. Extension services deter the adoption of organic agriculture because agents are trained to advise farmers to use what the experiment station has determined to be the most efficient inputs. Most often, organic methods are considered obsolete throwbacks to a less efficient time.

Land tenure is another determining factor, given the long-term commitment needed for organic methods to be effective. Tenant farmers are unlikely to invest the necessary labour and sustain the costly conversion period without a guarantee of continued access to the land. Organic farmers may take years to reap a return on their investment and tenants and sharecroppers seldom have the luxury to wait that long.

The trade of organic commodities needs to be viewed with reference to international commodity markets. Prices received by farmers for conventional products have stagnated or decreased in real terms over the past thirty years, with farmers sometimes collecting revenues below production costs. Relative decline in prices since the late 1990s affects almost all agricultural commodities. Low coffee prices have forced millions of small farmers into crippling debt that ultimately results in them forfeiting their land.

Low world prices also mean low export earnings: developing countries' export earnings from beverage crops fell by 18 percent between 1999-2000. Current low price levels of agricultural commodities are likely to remain so in the short-term. Costs of conventional farming inputs have increased substantially, usually requiring hard currency to import.

Organic agriculture offers an opportunity to improve income because of: consumers' willingness to pay price premiums for organic produce and lower production costs due to reduced use or absence of imported inputs. Even when the price premium on organic products is low, stable and profitable long-term prices offer more security to farmers than volatile conventional commodity markets.

Institutional Performance

Because "organic" is a production process claim, consumers must rely on certification programmes that verify claims. The standards that specify the organic production process appear quite precise, especially compared with claims made for other competing products. Consumers decide to purchase organic food in part because that choice reflects their values. For example, many consumers demand that organic food meet strict animal welfare standards. Others expect organic to mean fresh, local and minimally processed. The whole organic community requires that genetically modified organisms not be used to produce or process organic food or fibre. Social and economic issues, including producers receiving a fair price, are increasingly receiving attention.

Most industrialised countries have regulations that govern organic agriculture, including Australia and the European Union countries, Switzerland, Japan and the United States. Some developing countries have also established policies and regulations on organic agriculture. In February 2002, 56 countries were reported to be at some stage of regulating the organic sector. Differences with regards to the scope of regulations and variation in their implementation raise a number of concerns, namely:

- import discrimination whereby compliance is required with standards not always suitable to agro-ecological conditions of export countries;
- multiple accreditation of certification bodies in order to access the three main organic agriculture markets (Europe, Japan and the United States);
- difficulties to traders, due to different rules interpretation by certification bodies;
- enormous workload (and delays) for authorities to negotiate bilateral equivalency;
- limitation of bilateral agreements with regards to products with ingredients sourced from around the globe.

The Codex Alimentarius Guidelines for the Production, Processing, Labelling and Marketing of Organically Produced Foods constitute a recognised basis to harmonise organic standards and national regulations. The Codex Committee on Food Import and Export Inspection and Certification Systems is developing Draft Guidelines on the Judgement of Equivalence of Technical Regulations Associated with Foods Inspection and Certification Systems which intend to develop an infrastructure for the review of technical requirements other than sanitary measures associated with inspection and certification systems. These Guidelines suggest a process and general principles to determine the equivalence of all food systems and are particularly relevant to organic agriculture.

In the absence of official measures to address equivalence, the organic community has organised an international programme to accredit certification bodies. The IFOAM Accreditation Programme, established in 1992, developed international procedures to evaluate organic certification programmes and assess compliance of organisations that claim to adhere to organic standards. Certification programmes involved with organic agriculture apply to the International Organic Accreditation Service (IOAS), an NGO established in 1997, in order to be accredited. IOAS evaluates their standards against the IFOAM Basic Standards and examines the competence of their programme against established criteria through field visits and audits. To date, IOAS has accredited about 20 certification bodies, operating in both developed and developing countries. This privately organised service has the potential to facilitate international trade in organic products, but this promise will be fully realised only when it is recognised by governments which have developed rules for organic agriculture.

Developing countries are important suppliers of organic commodities. They, however, need to establish that they conform to the standards and rules of the importing developed countries. Suppliers to multiple markets may need to carry several such certifications: standards accepted in Sweden may not be recognised in the United States or Japan, and vice-versa. In countries where recognised domestic facilities are lacking, suppliers often must hire foreign inspection and certification bodies; in many instances, this is prohibitively expensive. Given that a certain part of certification is a fixed cost, certification costs take a higher percentage of earnings of smaller units. Smallholders in developing countries have little chance to export certified organic products without active government support for inspection and certification. Alternative control systems for small holders are however developing in order to ensure quality assurance without depending totally on foreign inspectors and certification bodies.

Many developing countries require external technical assistance to build the capacity for technical, organisational, and legal skills needed to establish reliable certification and accreditation programmes. Some certification bodies become accredited by the importing country. This requires educated, trained personnel and administrative structures. International equivalence of various national organic standards will reduce the administrative overhead, improve public-sector relations with private certifiers and traders, and eliminate redundant certification. This bureaucracy and its attendant cost particularly burdens poor farmers in developing countries. Internationally recognised accreditation and equivalence will benefit exporting and importing countries alike because it ensures conformity with requirements of importers while recognising the competence and compliance of the exporters.

Ensuring compliance must be legitimate and enforced with sufficient sanctions in both developed and developing countries alike. There is evidence that fraud exists in many developing countries. Recent experiences in Germany and in the United States show that developed countries are not free of negligence and fraud. This hurts the honest legitimate organic farmer in the exporting country as well as the consumer in the importing country. Trusted and enforced organic guarantee systems are key to successful trading of organic agriculture products. The establishment of an internationally agreed accreditation mechanism is crucial for determining equivalence of imported organic products.

ORGANIC AGRICULTURE AND THE ENVIRONMENT

In many countries, agriculture is the largest form of land use; farmland habitats account for a great percentage of natural habitats. Protected areas are not sufficient for nature conservation, especially for migratory species (e.g. birds) because surrounding habitats are agricultural lands which often create negative externalities. Agriculture, especially in its most extreme modern form of industrialised monoculture, modifies landscapes and damages the ecosystem's goods and services, including biodiversity at all levels. Agricultural encroachment, as well as pollution and intensification are responsible for contributing to soil and water degradation and to the extinction of biodiversity.

Current approaches to ecological agriculture such as integrated pest management, integrated plant nutrition systems and conservation tillage, consider only one aspect of the farming system components: pest ecology, plant ecology and soil ecology, respectively. The ability to define strategies that combine these and other management elements into a single approach is proper to organic agriculture. Organic management focuses on the food web relations and element cycling and aims to maximise the agro-ecosystem's stability and homeostasis.

Organic agriculture, through its systemic approach and avoidance of agro-chemicals, prevents natural resource degradation and the loss of land and productive potential. In organic agriculture, nature is both instrument and aim. As organic farmers cannot use synthetic substances (e.g. fertilisers, pesticides, pharmaceuticals) they need to restore the natural ecological balance because ecosystem functions are their main productive "input". For example:

— Many unspecific pests like aphids, thrips, whiteflies or spider mites, economically damaging in many crops, can be kept below the economic threshold with naturally occurring or purposely released predators and parasitoids. The former are direct goods and services of hedges, botanically diverse field margins, intercropping or weedy undergrowth, the latter do better when released in botanically and ecologically enriched habitats.

- The only way to suppress soil-borne pests and diseases in organic agriculture are wide crop rotations with several components of botanically different crops. Adhering to such rotations is crucial to providing agro-ecosystem diversity.
- Diverse crop rotations, or agro-forestry systems, guarantee a better uptake of nutrient elements from the soil and very efficient use of water and light, thanks to varying spatial and temporal root growth and leaf dispersion.
- Soils with a high functional diversity of micro-organisms, which occur very often after decades of organic agriculture practice, develop disease suppressive properties and can help to induce resistance in plants.
- By restricting farm inputs farmers have to use preventive techniques appropriately. The ban on herbicides, for instance, makes it impossible to ignore good crop rotation principles and disastrous in terms of yields and long-term problems with weeds. The ban on soluble and purchased fertilisers makes nutrient-conserving crop rotations and the sparing use of organic fertilisers to reduce losses, economically worthwhile.

As organic farmers cannot use synthetic substances (e.g. fertilisers, pesticides, pharmaceuticals) they need to restore the natural ecological balance because ecosystem functions are their main productive "input".

The direct economic value of ecosystem functions for organic farmers is an excellent example of "benefit-sharing". To master ecosystem functions in order to produce decent high-quality yields, adaptive management practices are crucial: growing conditions in organic agriculture are neither optimal, static nor predictable. Farmers must become excellent observers and must be trained to react flexibly, often intuitively and appropriately to the local context. In contrast, other farming methods try hard to condition the environment of the plants for optimal growing conditions by using a wide range of inputs. Moreover, most recommendations and fertilisation and spraying programmes are highly standardised and not site-specific.

Organic agriculture is an approach, which is not confined to production alone. Intersectoral cooperation between farmers, environmentalists, processors and traders, quality controllers and consumers is a very common characteristic of organic agriculture. Doherty et al. described sustainable agriculture as "ecological engineering solutions aiming to minimally manipulate and manage ecosystems for the benefit of both nature and humanity". To date, no ecological farming scheme other than the organic one has succeeded in providing ecosystem functions and socio-economic goods in a comparably equitable way.

Organic Agriculture and Biodiversity

For hundreds of years, agriculture has contributed substantially to the diversity of species and habitats, and agriculture has formed many of today's landscapes. Over the last century, however, modern intensive agriculture, with its high input of synthetic pesticides and fertilisers and monocrop specialisation, has been detrimental to the diversity of genetic resources of crop varieties and livestock breeds, to the diversity of wild flora and fauna species and to the diversity of ecosystems. The 2000 IUCN Red List of threatened species highlights habitat loss as the main threat to biodiversity, with agricultural activities affecting 70 percent of all threatened bird species and 49 percent of all plant species.

Because of the increasing loss of agricultural biodiversity on a global scale, the Convention on Biological Diversity has developed a work programme on this subject in 1996. This programme states inter alia that farming practices that stop degradation while restoring and enhancing biological diversity should be encouraged, including organic agriculture.

Organic agriculture is dependent upon stabilising agro-ecosystems, maintaining ecological balances, developing biological processes to their optimum, and linking agricultural activities with the conservation of biodiversity. Wild species perform a variety of ecological services within organic systems: pollination, pest control, maintenance of soil fertility. Thus, higher levels of biodiversity can strengthen functions essential for farming systems and therefore, agricultural performance. Enhancing functional biodiversity is a key ecological strategy to bring sustainability to production on organic farms. Organic systems also use substantially fewer external inputs and do not use synthetic chemical fertilisers, pesticides, genetic modified organisms and pharmaceuticals. Instead, systems are designed to work in harmony with nature in order to determine agricultural yields and disease resistance. By respecting the natural capacity of plants, animals and the landscape, organic agriculture aims to optimise quality in all aspects of agriculture and the environment.

Biological pest control on organic farms, for example, relies on maintaining healthy populations of pest predators and parasitoids. A study in California comparing conventional with organic tomato fields showed a higher abundance of natural enemies and greater richness of species in organic tomato fields. There was no significant difference for any type of damage to tomato foliage or fruit, showing that the organic system achieves the same levels of pest control without having to apply synthetic chemical pesticides.

Organic agriculture is thus committed to the conservation and enhancement of biodiversity within agricultural systems, both from a philosophical perspective and from the practical viewpoint of maintaining productivity. To this end, the importance of biodiversity as part of a well-balanced organic system is part of the IFOAM International Basic Standards for Organic Agriculture Production and Processing.

Numerous scientific studies, mainly from Europe and North America, give evidence that on organic farms biodiversity is higher than that on conventional and farms. Biodiversity is generally assessed at three distinct levels :

- Genetic diversity: the variation between individuals and between populations within a species;
- Species diversity: the different types of plants, animals and other life forms within a region or community;
- Ecosystem diversity: the variety of habitats found within an area (grassland, marsh, and woodland for instance). These levels of biodiversity have been used in the following sections to show the links between biodiversity and organic agriculture.

In general, the degree of biodiversity in agro-ecosystems depends on four main characteristics of agro-ecosystems :

- The diversity of vegetation within and around the agro-ecosystem.
- The permanence of various crops within the agro-ecosystem.
- The intensity of management/farming.
- The extent of isolation of the agro-ecosystem from natural vegetation.

Apart from the adoption of high-yielding, uniform cultivars, a further possible threat to genetic diversity, and biodiversity in general, are the side-effects of the release of genetically engineered organisms into the environment.

Genetically engineered plants designed to control pests can have negative side-effects on beneficial insects and further non-target organisms as well. Oilseed rape with genetically induced resistance to insects has been reported to damage beneficial insects such as honey bees. The use of herbicide resistant plants can result in greater use of herbicides, increasing the negative effects of intensive farming on natural biodiversity. Furthermore there is the danger that transgenic plants could become feral and thus suppress indigenous flora. Feral oilseed rape populations in Canada are resistant to three herbicides and have become one of the most troublesome weeds.

Recently, it was reported that transgenic DNA introgressed into traditional maize landraces in Oaxaca, Mexico. Concerns have been raised about the potential effects of transgenic introductions on the genetic diversity of crop landraces and wild relatives in areas of crop origin and diversification, as this diversity is considered essential for global food security.

Insofar as organic agriculture is dependent upon maintaining ecological balances and diverse agro-ecosystems, genetic engineering is a contradiction to the principal aims of organic agriculture. Organic agriculture does not allow genetic engineering in its standards as genetic engineering focuses on genetic makeup without taking into account the complete organism or farming system in which the organism functions.

Ecosystem Diversity

According to the World Resources Institute, an ecosystem is made up of the organisms of a particular habitat, such as a farm or forest, together with the physical landscape in which they live. Although little research has been carried out comparing agro-ecosystem diversity in different farming regimes, many of the principles of organic agriculture are likely to have a positive impact on ecosystem diversity.

The 2002 IFOAM Basic Standards for Organic Agriculture Production and Processing include principles and recommendations on "organic ecosystems" where provisions are made to "maintain a significant portion of farms to facilitate biodiversity and nature conservation", including (among others) wildlife refuge habitats and wildlife corridors that provide linkages and connectivity to native habitat.

The Swiss organic standards require organic farmers to use 7 percent of their land as semi-natural habitats, including field margins. Many other organic certification bodies have included biodiversity requirements in their standards. This applies to organic and low-input farmers receiving direct payments under the agri-environment schemes. Direct payments by the state are linked to ecological performance which is controlled by inspection services.

Organic agriculture has a high and possibly decisive potential for reversing the dramatic decline of biodiversity. In Sweden, a working group made up of organic farmer groups, nature conservationists, Government agencies and universities has been working since 1997 to strengthen the links between organic agriculture and biodiversity conservation, and they set up biodiversity plans for organic farms. The main objectives, apart from helping organic agriculture enhance biodiversity, have been to start cooperation and dialogue between the nature conservation and organic agricultural movements and to spread knowledge about biodiversity in organic agriculture. From there, the discussions developed into a planning exercise leading to proposals for changes to the country's main organic standard (set by the certification body, KRAV), to require all organic farmers to have a plan for the management of biodiversity on their farms from 2001.

Organic agriculture has a high and possibly decisive potential for reversing the dramatic decline of biodiversity. However, further efforts must be taken to refine the basic organic standards for biodiversity and landscape. Actual experiences show that there is a considerable and diverse synergy potential between biodiversity conservation and farm income. Biodiversity issues based on measurable performance are more and more important as a basis for subsidies and, at least in some countries, for consumer decisions for buying organic products.

Semi-natural habitats as part of organic agriculture

The conservation and the management of semi-natural habitats play an important role in organic agriculture. Semi-natural habitats are refuges for endangered plant species which in former times were found in meadows and arable fields. Semi-natural habitats and field margins are also essential for the survival of many invertebrates, especially due to favourable food and over wintering conditions. They also function as habitat cross-links between meadows, fallows and different field margins.

Quantitative data from Switzerland found that the proportion of semi-natural habitats and field margins per farm to be 16 percent on organic farms, compared to 3.7 percent on conventional farms. It was also noted that in Germany organic farmers took specific measures to increase the habitat diversity (hedges, low-input orchards, ponds, corridors, habitat networks, wildlife refuges and devices) as well as to better connect farmed areas and the surrounding habitats.

A recent study , also in Switzerland, shows how important the combination and integration of semi-natural habitats with organic farms is for the conservation and enhancement of species diversity and abundance of beneficial arthropods. In this study organic agriculture had been compared with low-input integrated crop management within the framework of a paired farm survey. Also nearby semi-natural landscapes have been used to compare the effects of organic agriculture and low-input farming system on carabid beetle and spider fauna. Many endangered or rare species which are enhanced by semi-natural habitats and field margins were more abundant in organic arable fields than in the integrated managed fields. Also, agro-ecologically important carabid species and wolfspiders were found in higher numbers on the organic farms. This aspect indicates that the improvement of landscape infrastructure in combination with organic agriculture may be an important factor for the conservation and enhancement of species-rich communities on agricultural land.

Landscape development and organic agriculture

Organic agriculture has the potential for a positive landscape development because it itself need a rich and diverse landscape infrastructure with semi-natural habitats. By studying the whole farm, or even the whole landscape within which the farm operates, researchers are trying to find ways to characterise the benefits of organic agriculture to the ecosystem. A research project funded by the European Union found that the diversity of landscape and farming systems was greater in organic farms, regarding land use types, crops, livestock, plantings (hedges, solitary shrubs, trees) and flora. In terms of landscape diversity, the organic types of agriculture have a good potential for positive contributions to a sustainable agrarian landscape.

Protected areas

Protected areas, as defined by the World Commission on Protected Areas, are areas especially dedicated to the protection and maintenance of biological diversity. Protected areas are not always strict nature reserves. Instead, they can fulfil many functions alongside biodiversity conservation. Because of its biodiversity benefits, organic agriculture offers an important agricultural management option in several protected area categories.

Because of its biodiversity benefits, organic agriculture offers an important agricultural management option in several protected area categories.

In Italy for instance, the Italian Association for Organic Agriculture (AIAB) project on "Organic Agriculture and Agro-ecology in Regional Parks" has been working with the regional park authorities in the Emilia-Romagna region to promote organic agriculture in relation to the regional agri-environment programme. During the first two years of activity there was an increase in the rate of adoption of the regional agri-environment programme, particularly of organic agriculture, by farmers in the park and the buffer zone. Between 1996 and 1997, 113 farms in the area applied for organic certification, compared with only 73 between 1994 and 1996.

There are further examples from Europe and the United States where organic agriculture is promoted in the biosphere reserves. The Estancia Itabo in Paraguay embodies a protected area of 5 000 hectares of high quality Interior Atlantic Forest. The Estancia is a good example of sustainable rainforest use in line with the conservation goals of the protected area by cultivating organic yerba mate and the heart of palm Euterpe edulis.

Support for Organic Agriculture for its Biodiversity Benefits

There is increasing recognition among nature conservation organisations that many species interact with agricultural systems, even if their primary habitat is in natural areas. The management of these agricultural systems can, thus, dramatically affect overall levels of biodiversity, as well as the success of particular species. The German conservation organisation BUND claims that traditional concepts of nature protection have failed because conservation goals have not been achieved, even in protected areas, due to agricultural intensification in the surrounding areas. Many conservation organisations are therefore calling for a general extensification of agricultural land use and promote organic agriculture for that reason.

In 1999, a joint workshop in Vignola, Italy, organised by The World Conservation Union (IUCN) and IFOAM together with the World Wide Fund for Nature (WWF), was held to exchange ideas and information on organic agriculture and biodiversity. A joint action plan for both the nature conservation and organic movements was drawn up in the so-called Vignola Declaration. In both developed and developing countries, one of the driving forces behind the growth of organic agriculture is the nature conservation agenda. Though less significant than market prospects and development prospects, a recent literature survey highlights a number of examples where nature conservation organisations are working closely with local farmers who live in, or close to, areas of significant nature conservation interest. In the last twenty years, governments also have been gradually moving towards policies aiming at encouraging the links between organic agriculture and biodiversity conservation.

In the European Union, environmental concerns have gained a more prominent place in agricultural policy. The most notable contribution of the Common Agricultural Policy of the European Union towards more environmentally sustainable systems has been the introduction of the agri-environment programmes in 1993, which now continue under Agenda 2000. The impetus for some of these initiatives comes from the Convention on Biological Diversity.

The implementation of agri-environmental measures over the whole of the European Union and also other European countries is the core of the European Community's environmental strategy. So far, organic agriculture has played a central role in most countries' national agri-environment policy. The main reason for this policy support has been the perceived positive environmental effects of organic agriculture.Also in other parts of the world, governments and development agencies are beginning to promote organic agriculture, especially in protected areas, because of its biodiversity benefits.

Agricultural-related environmental problems relate chiefly to water quality and intensive livestock operations. To illustrate this point, research using the United Kingdom Government data suggests that the costs of meeting the Nitrates Directive in England and Wales reduced net farm income by 2.4 percent in the poultry sector, 11.3 percent in the pig industry, 4.5 for beef and 9.6 on dairy farms. Most of the cost relates to capital expenditure to adequately handle manure. Animal welfare regulations for poultry and pigs are estimated to reduce net farm income by 38 and 6 percent respectively. Crops do not face the same compliance costs. The implications of these estimates are that perhaps environmental concerns do not justify the abandonment of pesticides and synthetic fertilisers that a widespread shift to organic agriculture would entail. However, this approach ignores the environmental damage that occurs even though the regulations are being met.

There is some evidence that consumers who have recently entered the organic market are less motivated by ecological issues and are more interested by organic food as a safe product. What is interesting is that sales of organic products have increased markedly as a result of recent food scares, suggesting that new consumers are concerned about the product itself as well as, or perhaps rather than, the environment in which it is produced. However, the increases in sales have not related to the particular products (that is, beef) to which the scares applied. This distinction between food safety and environmental concerns has important implications for future production and policies.

ORGANIC AQUACULTURE – FUTURE PROSPECTS

Aquaculture (the farming of aquatic animals and plants), much like organic agriculture, is one of the world's fastest growing food sectors. Globally, aquaculture production has been growing at an average rate of 9 percent per year since 1970, compared with 2.9 percent for terrestrial farmed meat production and 1.3 percent for capture fisheries. However, to date, aquaculture has lagged behind the agriculture sector in terms of the quantities and diversity of certified organic produce.

This delay is largely due to the absence of universally accepted standards and accreditation criteria for the production of organic aquaculture produce. It is also due to certifying bodies being almost totally restricted (until very recently) to a handful of organisations within developed countries (Europe, North America and Oceania) even though these countries produced less than 10 percent of total global aquaculture production in 1999. Aquaculture has lagged behind the agriculture sector in terms of the quantities and diversity of certified organic produce.

Although no official statistical data are available concerning the global production of certified organic aquaculture products, it is estimated that total production in 2000 was only about 5 000 metric tonnes, primarily from European countries. This modest quantity represents about 0.01 percent of total global aquaculture production or about 0.25 percent of total European aquaculture production. According to Bergleiter the total volume of organic aquaculture products marketed in Europe in 2000 was between 4 400 and 4 700 tonnes. These included:

- 4 000 tonnes of salmon (produced from Irish and Scottish farms for sale to Austria, Belgium, France, Germany, Ireland, Luxembourg, the Netherlands, Switzerland and the United Kingdom);
- 100-200 tonnes of trout (produced from Scottish and German farms for sale to local domestic markets);

- 200-400 tonnes of carp and accompanying freshwater species such as tench (produced from Austrian and German farms for sale mainly to domestic markets);
- 100 tonnes of blue mussels (produced from one Irish farm for sale in Germany).

Unfortunately, little or no production data is available for countries outside Europe. New Zealand was one of the largest producers outside Europe. A salmon farm was the first certified aquaculture facility in New Zealand with an estimated 500-800 tonnes of organic salmon targeted for the European market (the production of which has since been discontinued). Non-European countries which are actively trying to develop their organic aquaculture production industries using national or private standards include Australia, Canada (salmonids), Chile (salmonids), Ecuador (shrimp), Indonesia (shrimp), New Zealand (mussels), Peru (shrimp), Thailand (shrimp), Viet Nam (shrimp) and the United States (non-species specific).

The slow initial growth of organic aquaculture has been due the absence of internationally recognised and universally accepted regulations and standards for producing and handling organic aquaculture products. For example, although both the FAO/WHO Codex Alimentarius Commission and the European Union have produced guidelines and standards for organically produced foods, neither have yet dealt with organic aquaculture.

The slow initial growth of organic aquaculture has been due the absence of internationally recognised and universally accepted regulations and standards for producing and handling organic aquaculture products. Realising the need to address this issue, the International Federation of Organic Agriculture Movements (IFOAM) drafted Basic Standards for Organic Aquaculture Production. These guidelines were first prepared in 1998 and adopted as draft standards by IFOAM at its General Assembly in Basel, Switzerland, in 2000. The draft standards have been subsequently further revised (based on the deliberations and inputs of a IFOAM Organic Aquaculture Working Group and from inputs received from other interested stakeholders) and are expected to be finally voted into full standards at the next General Assembly of IFOAM.

The United States is also making progress, albeit at a slower pace. Ten years after the United States Organic Food Production Act was promulgated, the United States Department of Agriculture's National Organic Program (USDA/NOP) published the long-awaited Final Rule on National Standards for Organic Crop and Livestock Production, Handling and Processing. Aquatic animal standards are not yet included; however, they will eventually be amended into the Final Rule. The National Organic Standards Board (NOSB) established an Aquatic Task Force and Aquaculture Working Group in 2000 to examine key issues and formulate recommendations for submission to the USDA/NOP. These were submitted in October 2001. Of the recommendations, the maximum allowance of 5 percent fishmeal and oil in aquatic animal diets is perhaps the most significant and constraining for United States organic aquaculture producers. This is due to the existing National Standards mandating that livestock under organic management be provided 100 percent organic feed. This is also applicable to organic aquatic animals; therefore, fishmeal and oil, required by many aquatic animals, must also be organic. Despite the intense activity of the Aquatic Task Force and Working Group, it remains unclear when exactly the USDA/NOP will draft a proposed organic aquaculture rule. In addition, the Organic Food Production Act specifically states that "on or after October 1, 1993 a person may sell or label an agricultural product as organically produced only if such product is produced and handled in accordance with this chapter." This leaves aquaculturists who are considering transitioning into organic production, and those that have already earned private organic certification, in a difficult position, especially as the October 2002 Final Rule implementation date approaches. As of December 2001, The USDA is still undecided about whether it will grant these producers the much-needed opportunity to call their product organic after October 2002.

In the absence of these international/regional standards, it has been left to individual member states and private/non-governmental certifying agents to set and develop their own specific organic aquaculture standards and accreditation bodies. For example, although the Soil Association produced its first draft organic aquaculture standards as early as 1989, it was not until 1998 that subsequent revisions of these draft standards were eventually approved by their Council as interim standards, with certified organic salmon and trout reaching the marketplace the following year (the interim standards only applying to salmonid production). These salmonid aquaculture standards, together with those of two other United Kingdom private certifiers (Food Certification Scotland Ltd and the Organic Food Federation) have since been officially recognised, in accordance with European Union Council Regulation 2092/91, by the United Kingdom Register of Organic Food Standards their official Standards for Organic Food Production. To date the only other country that has developed national aquaculture standards has been France and more recently, in September 2001, by Australia.

A pioneering private certifying body actively engaged in the promotion and development of organic aquaculture in Europe and globally is Naturland, based in Germany. The association launched their activities in 1995 with development of organic aquaculture standards and initial certification of organic carp and tench production in Southern Germany using traditional pond culture techniques. The development of standards for salmonid and mussel farmers in Ireland followed a few years later with organic salmon and mussels reaching markets in 1996 and 1999, respectively. More recently, Naturland extended its reach to developing countries by developing standards for the organic production of shrimp in ponds.

Farmed shrimp represents the single most valuable internationally traded aquaculture commodity worldwide, with 99.4 percent of total global production originating from developing countries within the Asia and Latin America Regions. With the support of German GTZ (Gesellschaft für Technische Zusammenarbeit mbH), Naturland initiated their first pilot project for the organic production of shrimp in Ecuador, and since then other countries (including Indonesia, Peru and Viet Nam) have shown interest in the project.

Despite its late start and modest size, the organic aquaculture sector currently boasts 20-25 private and non-private certifying bodies. They have a diverse set of aquaculture standards which sometimes vary considerably from country to country, certifier to certifier, and species to species. To a large extent these differences reflect the differences between individual certifiers, farmers, and other interested stakeholders in the interpretation of what organic aquaculture really means and entails. There is an urgent need for the universal acceptance and adoption of a broad set of basic principles and production standards.

Organic Aquaculture and the Environment

In marked contrast to the freshwater-dependent terrestrial agricultural production systems, aquaculture can also be realised within marine and/or brackish water environments. For example, over half (54.7 percent) of total global aquaculture production currently originates from marine or brackish coastal waters. This includes aquatic plants and molluscs within marine waters and crustaceans (shrimp, crabs) and finfish (mainly salmonids) in brackish water. In the case of the total reported certified organic aquaculture products produced in Europe, 87-93 percent of these were produced in marine and brackish waters. The use of these hitherto largely untapped vast aquatic resources is particularly essential in view of the urgent need to conserve our precious fresh water supplies for human consumption and conventional agriculture, including livestock production. In addition to organic fish and mollusc production, the seas hold particular promise for the production of organic aquatic plants for either for direct human consumption or as much needed organic feed inputs for animal husbandry.

For the organic aquaculture sector to successfully co-exist with other food production sectors, it will have to successfully source its own organic feed and nutrient resources. For the organic aquaculture sector to successfully co-exist with other food production sectors, it will have to successfully source its own organic feed and nutrient resources. For example, a major concern with the organic production of carnivorous fish species such as salmon and trout (over 73 percent of farmed finfish production within developed countries currently being carnivorous finfish species) is the use or not of fish meal and fish oil within organic feeds for these species. In particular, questions revolve around:

— whether a product derived from wild caught animals can be certified ;

— what the maximum level of fish meal or fish oil is that can be used within certified organic feeds ;

— the transfer of essential protein and lipid sources from one part of the globe to the other ;

— concerning the ethics and long term sustainability of producing organic carnivorous fish species.

Clearly, if organic principles are to be upheld, it is essential that these products be:

— obtained from sustainably-managed fisheries (according to internationally accepted management guidelines);

— derived from locally-available fishery products (including fish processing waste) not suitable for direct human consumption;

— free from synthetic additives and unwanted contaminants;

— only fed to farmed organic aquatic species with naturally piscivorous feeding habits.

Utilising public water bodies for aquacultural production is not necessarily a silver bullet for organic production, particularly for culture of animals reared in net cages.

Utilising public water bodies for aquacultural production is not necessarily a silver bullet for organic production, particularly for culture of animals reared in net cages. The vulnerability of net-cages in open, aquatic ecosystems brings a unique set of problems including an increase in escapes from these systems, untreated release of effluents and exposure to chemical drift. Although these problems are not necessarily barriers, they can be considered limitations to organic certification by some standardising bodies. Although this issue may be particularly contentious in the United States, within Europe other issues such as organic fish being able to perform their "natural behaviour" (i.e. roam freely and exhibit normal migration behaviour) could become equally as important. It is, however, important to remember here that over 91 percent of total conventional marine aquaculture production in 1999 was farmed aquatic plant and mollusc species feeding low on the aquatic food chain. As mentioned previously, with the possible introduction of appropriate water and nutrient management techniques, the prospects for the increased production of farmed organic aquatic plants and molluscs is considerable.

REFERENCES

Anon. 1998. "If you care about organic food...act now!". *Organic Gardening*. March. p. 22–25.

Hainsworth, P.H. 1976. Agriculture: The Only Right Approach. Bargyla & Gylver Rateaver, Pauma Valley, CA . p. 27.

Hudson, Berman E. 1994. "Soil organic matter and available water capacity". *Journal of Soil and Water Conservation*. Volume 49, Number 2. p. 189-194.

Lokeretz, William, Georgia Shearer, and Daniel Kohl. 1981. "Organic farming in the Corn Belt". *Science*. Vol. 211, No. 6. February.

Magdoff, F.R. 1992. *Building soils for better crops: organic matter management*. University of Nebraska Press, NE.

Pretty, J.N. 1994. *Regenerating Agriculture*. Earthscan Publications Ltd., London.

Reijntjes, C.B., Haverkort and A. Waters-Bayer. 1992. *Farming for the future*. MacMillan Press Ltd., London.

Savory, Allan, with Jody Butterfield. 1999. *Holistic Management*. Island Press. Washington, D.C. 616 p.

Sumner, D.R. 1982. "Crop rotation and plant productivity". In M. Recheigl, (ed). CRC *Handbook of Agricultural Productivity*, Vol. I CRC Press, Florida.

USDA Study Team on Organic Farming. 1980. *Report and Recommendations on Organic Farming*. US Gov't Printing Office 1980-0-310-944/96. USDA, Beltsville , MD. July.

5

Holistic Farm Management

Sustainable agriculture seeks in principle to "sustain" economic viability, environmental stewardship, and social responsibility. These three tenets are to be embraced as one functional unit. Decisions concerning a sustainable agriculture should then enhance the environment and the farmers economic situation and benefit the regional society. Holistic Management gives us a way to move forward on these three tenets. It gives us a way to design agriculture to truly mimic nature's principles of sustainability. It gives us a way to make decisions that automatically take into account the society, the economics, and the environment before they are made.

Holistic management is a simple decision-making framework that can be learned like any other skill. People who manage holistically can realise an improved quality of life and generate real wealth, while at the same time improving the land and community around them. They develop the ability to ask the right questions and to confidently proceed toward the future they design for themselves. Holistic Management is for anyone who wants consistent profit from agriculture, a high quality of life, and more time to enjoy it. In short, it's a way to have fun, make money and conserve our natural resource base, all at the same time. Holistic Management is a process for sorting out and making sense of all the tools and choices that face us each day. Once a person begins to manage holistically, he or she learns what to say no to, and what to say yes to. Being a *proactive* process, holistic managers learn how to move beyond crisis management and toward planned prosperity. They are able to manage their finances wisely, have more time for enjoyment, live life according to their values, and gain the confidence of knowing that their decisions are improving the environment and the community they live in leading to a better world for their grandchildren.

HOLISTIC MANAGEMENT PROCESS

As the name "holistic" implies, resources are managed in whole units rather than as parts in isolation from their surroundings. In order to have a clear description of what is being managed, people begin by defining their whole. This definition involves a listing of all the decision-makers involved in management, the resources they have to work with, and the money available. From there, a detailed holistic goal is developed.

The holistic goal includes a values-based quality of life statement, a listing of forms of production that will make the quality of life possible, and a description of how the land base needs to be far into the future, in order to sustain the production. Defining the whole and writing a goal is a powerful exercise. People who have written goals are much more likely to succeed than those who do not. Since the holistic goal is based on the deeper underlying values of the decision-makers, it empowers them to ask better questions, to ask the deeper questions, to ask appropriate questions from which they can make better decisions.

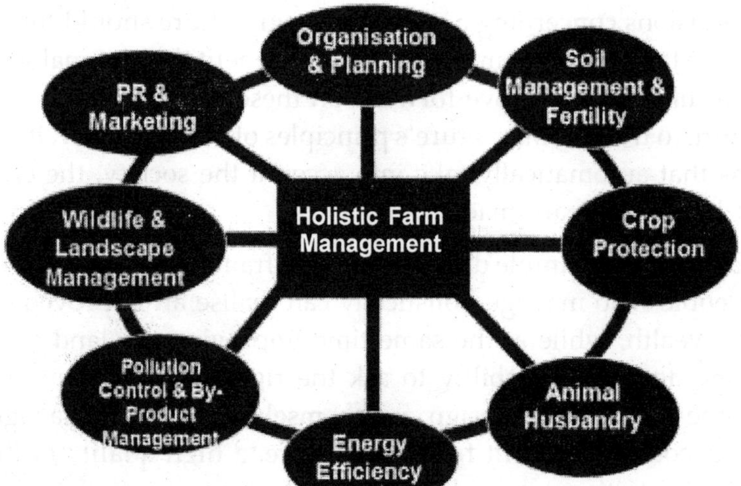

Figure 1. Holistic Farm Management

The holistic goal remains the centerpiece of holistic management and is referred to constantly when management decisions are being made. The goal is what drives the decision-making. But there's more. In order to sustain a farm operation, profit must come from somewhere. Most likely, at least some of the profit will come from on-farm enterprises.

Financial Planning

Holistic managers use a potent financial planning process that empowers them to make decisions that are simultaneously good for the environment, the local community and the bottom line. The holistic financial plan provides a road map to help people navigate through their financial year, assured that the profit will be there at year's end.

The financial plan allows managers to select enterprises that do not conflict with their values, and then to plan a profit up front. Once the profit is planned from the expected income, expense dollars are allocated sequentially where they will do the most good. Holistic financial planning differs from conventional financial planning in several ways. Conventional cash flow budgeting involves estimating income from an enterprise, then allocating expenses for capital investment, variable costs, and fixed costs. Attempts are made to keep costs below anticipated gross income by using past records and other information and adjusting for cost trends.

As long as the expenses appear cost effective and the plan predicts no cash shortages the bank won't cover, all should go well. Still, the results of cash flow planning often include considerable anxiety towards year's end over the profit margin.

In many cases, the expenses nearly equal the planned gross income, producing very little profit. Usually there are plenty of excuses to make up for the small margin weather, markets, and pests. All too often we may take the attitude of, oh well, things will be better next year. With holistic financial planning the projected income is planned, then the desired profit is allocated at the outset, heavily affecting how the remainder of the budget will be allocated. Planning a hefty profit before any expenses are allocated is a key distinction.

After profit is planned, expenses are allocated into three categories: Wealth generating, Inescapable, and Maintenance (WIM). Wealth generating expenses produce profit for the operation this year. Inescapable expenses must be paid regardless (taxes, land payments, etc.), while maintenance expenses, though essential to the business, do not produce profit this year.

Overhead and variable categories are not used at all because they don't describe what the expenses within each category do for the enterprise. Once the holistic financial plan is written, it is monitored monthly to stay on track toward the planned profit. Monthly monitoring allows deviations from plan to be caught early and corrected before there is serious financial trouble.

The financial planning process helps control three human tendencies that work against financial success:

1) the tendency to allow cost of production to rise to the level of optimistically anticipated income,

2) the tendency to borrow heavily against the optimistically anticipated income,

3) the tendency to do little planning ahead of time on paper.

Even when planning is done using conventional cash flow budgeting, production is the goal, not profit. With holistic financial planning, profit is the goal and production is the means of achieving it. That's why profit is allocated right off the top of anticipated income. Planning a hefty profit up front forces the manager to overcome the three tendencies that lead to low profit margins.

From there, meeting necessary expenses after planning our profit requires creativity. How the profit is used at the end of the year is unimportant, but the objective of holistic planning is to make sure there will be a substantial profit at the end of the year.

Some of the key distinctions between holistic financial planning and the conventional cash flow model that most farmers use are shown in Table 1.

Table 1. Distinctions between holistic financial planning and cash flow planning.

Holistic	Cash-flow
Profit is the goal	Production is the goal
Profit is planned first	Profit is what's left over
Expenses put into WIM categories	Expenses put in overhead and variable costs
Monthly monitoring to stay on track	Annual monitoring?

With the holistic budget, the expenses are put into categories of wealth generating expenses, inescapable expenses, and maintenance expenses. Using these categories, one can easily see where to cut expenses, while preserving our planned profit and still meeting our debt obligations. The excess maintenance expenses of $26,000 will have to be cut by creative means to preserve our profit and still meet the debt obligation. There is much more to holistic financial planning than has been introduced here. Some additional aspects include managing debt, testing financial decisions toward a well-defined goal, creating and using a livestock production worksheet to plan cattle buying and selling, and brainstorming new enterprises.

Landscape Management

Since we all depend directly on the landscape for our very existence (food, clothes, water, etc.), we benefit greatly from gaining a complete understanding of how the landscape functions. The very essence of the term "holistic" is that nature functions only in wholes, not in parts, and that we will understand nature better when we manage it as a whole rather than as separate parts.

Holistic Management gives people a way to make decisions that more accurately mirror the way nature functions (in wholes) and thereby ensure that our farming is truly sustainable over time. To better understand how nature functions, her basic processes need to be considered. Looking closely, four basic processes can be found in all natural systems.

First, water falls to earth as rain, filters through the soil and is either taken up by plants or continues downward to become ground water. When water is cycling effectively, floods are infrequent and of lower impact, water is released slowly through underground flow into springs and streams, and erosion is virtually non-existent. If on the other hand, bare soil is exposed and plant density is low, most water runs off the landscape rapidly resulting in soil erosion, much less water entry into the soil, and severe and more frequent flooding. So, an effective water cycle is apparent in nature and essential to a sustainable agriculture.

A second natural process we can observe in nature is the mineral cycle through the biological system. Minerals needed for biological growth are constantly recycled from soil to plant to animal and back to soil again. There is very little waste in the natural mineral cycle. There is no need for fertiliser in nature, as all the fertility is recycled again and again with very little loss. Ultimately, to be sustainable, we need to find ways to utilise the natural mineral cycle while minimising our off-farm purchase of minerals.

Farming practices that inhibit the natural mineral cycle, only reduce the sustainability of our farm. A third natural process shows us that plant and animal communities strive toward high biodiversity. Not only is diversity high in the numbers of species, but also the genetic diversity within species, and a wide age structure of each population present.

Greater diversity produces greater stability within the system. It also assures minimal pest problems. Large expanses of monoculture represent a simple level of diversity. Monocultures are almost never present in nature. Monocultures require great energy expenditure, either with fossil fuels or animal and human power to maintain.

Weed invasion is nature's way of injecting diversity into monocultural cropland. When biodiversity is increased, the cost of pest control and fertiliser is decreased. Crop rotation is the first step toward increasing biodiversity on the farm. It helps break weed and pest life cycles and provides complementary fertilisation to crops in sequence with each other. Advancing from rotation to strip intercrops represents an even higher level of biodiversity.

Strip intercrops of corn and soybeans or cotton and alfalfa are two examples. Increasing habitat for more beneficial organisms with more borders, windbreaks, and special plantings for natural enemies of pests represent even higher levels of biodiversity and stability.

The fourth natural process involves the flow of energy from the sun through the biological system. The sun is the fuel driving the biology of our farm. Energy flows from the sun through the ecosystem from one level to the next. Sunlight is absorbed by the green plant, enabling it to grow. Plants are eaten by animals that are in turn eaten by predators which are eaten by even higher predators. During each step, energy is being transferred from one level to the next. Energy is transferred below ground through plant roots that eventually die. The dead roots become food for decomposer organisms.

The waste and by-products from the primary decomposers are consumed by another set of secondary decomposers. Finally the residue is broken down into plant available nutrients and soil humus. At each step of the decomposition process, energy is either transferred from one organism to another or is lost as heat.

High energy flow is typified by a thick stand of green plants covering the soil for as long a time as possible. Growing mixtures of two or more plant types increases the leaf area available to capture sunlight. The volume of plants (tight spacing) also enhances energy flow. By growing two or more crops per year, we can lengthen the time that plants are in the field collecting solar energy. If soils are left bare, no sunlight is being converted into energy. When energy flow is reduced by periods when the soil is bare or without a crop, the decomposer organisms living in the soil are on a starvation diet.

When we modify any one of these natural processes (water cycle, mineral cycle, biodiversity, and energy flow) we affect the others as well after all, they function as a whole. When we build our farm enterprises around these natural processes, we have a plan that will sustain our family today and future generations tomorrow. After all, these are nature's rules.

The sooner we live by them rather than fighting them, the sooner we will produce a sustainable farm. When we fight nature's rules, we only hurt ourselves in the end.

HOLISTIC DECISION MAKING

The word "tools" is used broadly in holistic management. Though we tend to think first of technology in all its many forms when we think of tools, there are several other tools available to us.

The additional tools include: fire, rest, grazing, animal impact and living organisms. Three additional tools we may not consider as tools are money, labour, and creativity. These last three tools cannot be used alone but only in conjunction with other tools. Each of the above mentioned tools affects the landscape depending on when and how they are used and in what climatic region. In moist regions with frequent rainfall, rest restores biodiversity to natural landscapes. In dryer areas with seasonal rainfall, rest reduces biodiversity. In those drier regions, animal impact is most beneficial in restoring rangeland health.

Many of the technology tools we often use can be replaced by living organisms in creative ways. One example comes from a Canadian rancher who solved his gopher problem by erecting hawk perches over his pastures. When the hawks patrolled the area regularly by using the perches, the gophers left in search of safer feeding areas.

Testing Decisions

Decision making is handled in an organised fashion in Holistic Management. Each decision is subjected to several simple testing questions that enable the decision-maker to see the likely effects of that decision on the whole. By quickly running a decision through the testing questions you get some assurance that the decision will be sound environmentally, economically, and socially.

There are seven tests but not all will apply to every decision. If information is lacking to make the decision, the testing will catch it. Testing forces the manager to consider much more than just cost or gut feel. If the decision fails one or more tests, the decision may be modified and run back through the testing guidelines again. After a second testing failure the decision might be abandoned all together. After a person gains experience, the testing questions become internalised. From that point on, appropriate testing guidelines automatically come to mind when faced with decisions.

One way to look at testing decisions in this manner is to consider testing as the needle on a compass and the holistic goal as the magnetic north the compass is attracted to. All testing is done toward the holistic goal. In fact, the first question a person should ask is: does this decision take me closer to my holistic goal? If the answer is clearly no, then drop the decision. If the answer is yes or maybe, then test the decision further.

Monitoring Decisions

Because nature is so complex that we can only begin to understand it, decisions affecting the landscape are assumed wrong and closely monitored for early warning indicators of need for change. If range burning is used to increase the plant density, one would look at the plant density later to determine if the effort was successful. By assuming the decision is wrong, we humble ourselves to the great complexity in nature, thus forcing us to monitor to keep on track toward our holistic goal. If we assume our decision is right, we might not monitor at all, or if we did, it would be only to record the results. Once this new holistic perspective is internalised and decisions are made accordingly, things begin to change for the better. Profits increase, the environment improves, and rural families prosper.

REFERENCES

Dore, J. 1997. *Sustainability Indicators for Agriculture: Introductory Guide to Regional/National and On-farm Indicators*, Rural Industries Research and Development Corporation, Australia.

Myers, Mary. 1998. "Grass, legumes and a reverence for soil fertility allow small acreage Amish farms to thrive in today's economy". *The Stockman Grass Farmer*. June. p. 1-4.

Netting, Robert McC. 1993. *Smallholders, Householders: Farm Families and the Ecology of Intensive, Sustainable Agriculture*. Stanford Univ. Press, Palo Alto.

Reijntjes, C.B., Haverkort and A. Waters-Bayer. 1992. *Farming for the future*. MacMillan Press Ltd., London.

Savory, Allan, with Jody Butterfield. 1999. *Holistic Management*. Island Press. Washington, D.C. 616 p.

Thompson, Dick. 1997. *Alternatives in Agriculture*: 1996 Report. Thompson On-Farm Research and the Wallace Institute. p. 3-4.

6

Agriculture and Climate Change

The Earth's average surface temperature increased 1.3 degrees Fahrenheit over the past century, and is projected by the Intergovernmental Panel on Climate Change to increase by an additional 3.2 to 7.2 degrees over the 21st century. These seemingly slight changes in temperature could have profound implications for farmers and ranchers. According to the Environmental Protection Agency, an increase in average temperature can:

- lengthen the growing season in regions with relatively cool spring and fall seasons;
- adversely affect crops in regions where summer heat already limits production;
- increase soil evaporation rates; and increase the chances of severe droughts.

Innovative farming practices such as conservation tillage, organic production, improved cropping systems, land restoration, land use change and irrigation and water management, are ways that farmers can address climate change. Good management practices have multiple benefits that may also enhance profitability, improve farm energy efficiency and boost air and soil quality.

GLOBAL CLIMATE CHANGE

Natural shifts in global temperatures have occurred throughout human history. The 20th century, however, has seen a rapid rise in global temperatures. Scientists attribute the temp increase to a rise in carbon dioxide and other greenhouse gases released from the burning of fossil fuels, deforestation, agriculture and other industrial processes.

This phenomenon as the enhanced greenhouse effect. The naturally occurring greenhouse effect traps the heat of the sun before it can be released back into space. This allows the Earth's surface to remain warm and habitable. Increased levels of greenhouse gases enhance the naturally occurring greenhouse effect by trapping even more of the sun's heat, resulting in a global warming effect.

The primary greenhouse gases associated with agriculture are carbon dioxide (CO_2), methane (CH_4) and nitrous oxide (N_2O). Although carbon dioxide is the most prevalent greenhouse gas in the atmosphere, nitrous oxide and methane have longer durations in the atmosphere and absorb more long-wave radiation. Therefore, small quantities of methane and nitrous oxide can have significant effects on climate change. Several excellent resources and fact sheets explain the greenhouse effect and the science behind climate change.

Impact of Climate Change on Agriculture

Climate change may have beneficial as well as detrimental consequences for agriculture. Some research indicates that warmer temperatures lengthen growing seasons and increased carbon dioxide in the air results in higher yields from some crops. A warming climate and decreasing soil moisture can also result in production patterns shifting northward and an increasing need for irrigation.

Changes, however, will likely vary significantly by region. Geography will play a large role in how agriculture might benefit from climate change. While projections look favourable for some areas, the potential of increased climate variability and extremes are not necessarily considered. Benefits to agriculture might be offset by an increased likelihood of heat waves, drought, severe thunderstorms and tornadoes. An increase in climate variability makes adaptation difficult for farmers. The U.S. Department of Agriculture released a report in May 2008 that focused on the effects of climate on agriculture, specifically on cropping systems, pasture and grazing lands and animal management. The following findings are:

- With increased carbon dioxide and higher temperatures, the life cycle of grain and oilseed crops will likely progress more rapidly.
- The marketable yield of many horticultural crops, such as tomatoes, onions and fruits, is very likely to be more sensitive to climate change than grain and oilseed crops.
- Climate change is likely to lead to a northern migration of weeds. Many weeds respond more positively to increasing carbon dioxide than most cash crops.
- Disease pressure on crops and domestic animals will likely increase with earlier springs and warmer winters.
- Projected increases in temperature and a lengthening of the growing season will likely extend forage production into late fall and early spring.
- Climate change-induced shifts in plant species are already under way in rangelands. The establishment of perennial herbaceous species is reducing soil water availability early in the growing season.

- Higher temperatures will very likely reduce livestock production during the summer season, but these losses will be partially offset by warmer temperatures during the winter season.

Impact of Agriculture on Climate Change

Agriculture activities serve as both sources and sinks for greenhouse gases. Agriculture sinks of greenhouse gases are reservoirs of carbon that have been removed from the atmosphere through the process of biological carbon sequestration. The primary sources of greenhouse gases in agriculture are the production of nitrogenbased fertilisers; the combustion of fossil fuels such as coal, gasoline, diesel fuel and natural gas; and waste management. Livestock enteric fermentation, or the fermentation that takes place in the digestive systems of ruminant animals, results in methane emissions.

Carbon dioxide is removed from the atmosphere and converted to organic carbon through the process of photosynthesis. As organic carbon decomposes, it is converted back to carbon dioxide through the process of respiration. Conservation tillage, organic production, cover cropping and crop rotations can drastically increase the amount of carbon stored in soils. In 2005, agriculture accounted for from 10 to 12 percent of total global human caused emissions of greenhouse gases, according the Intergovernmental Panel on Climate Change.

In the United States, greenhouse gas emissions from agriculture account for 8 percent of all emissions and have increased since 1990. The following is evident from the information is:

- Despite some improvement in certain areas since 1990, the U.S. agricultural production sector increased its greenhouse gas emissions and expanded its role in climate change.
- The U.S. agricultural production sector is a net emitter of greenhouse gas emissions. That is, agricultural production annually creates more greenhouse gas emissions than it captures, despite the potential for the sector to sequester higher levels of carbon with management changes.
- The U.S. agricultural production sector contributes more greenhouse gas emissions from methane (CH_4) and nitrous oxide (N_2O) than from carbon dioxide (CO_2).
- Agricultural soil management is the single greatest contributor to greenhouse gas emissions from the U.S agricultural production sector.

Carbon Sequestration in Agriculture

Carbon sequestration in the agriculture sector refers to the capacity of agriculture lands and forests to remove carbon dioxide from the atmosphere. Carbon dioxide is absorbed by trees, plants and crops through photosynthesis and stored as carbon in biomass in tree trunks, branches, foliage and roots and soils.

Forests and stable grasslands are referred to as carbon sinks because they can store large amounts of carbon in their vegetation and root systems for long periods of time. Soils are the largest terrestrial sink for carbon on the planet. The ability of agriculture lands to store or sequester carbon depends on several factors, including climate, soil type, type of crop or vegetation cover and management practices.

The amount of carbon stored in soil organic matter is influenced by the addition of carbon from dead plant material and carbon losses from respiration, the decomposition process and both natural and human disturbance of the soil. By employing farming practices that involve minimal disturbance of the soil and encourage carbon sequestration, farmers may be able to slow or even reverse the loss of carbon from their fields. In the United States, forest and croplands currently sequester the equivalent of 12 percent of U.S. carbon dioxide emissions from the energy, transportation and industrial sectors.

ROLE OF AGRICULTURE IN CLIMATE CHANGE MITIGATION

Several farming practices and technologies can reduce greenhouse gas emissions and prevent climate change by enhancing carbon storage in soils; preserving existing soil carbon; and reducing carbon dioxide, methane and nitrous oxide emissions.

Conservation Tillage and Cover Crops

Conservation tillage refers to a number of strategies and techniques for establishing crops in the residue of previous crops, which are purposely left on the soil surface. Reducing tillage reduces soil disturbance and helps mitigate the release of soil carbon into the atmosphere. Conservation tillage also improves the carbon sequestration capacity of the soil. Additional benefits of conservation tillage include improved water conservation, reduced soil erosion, reduced fuel consumption, reduced compaction, increased planting and harvesting flexibility, reduced labour requirements and improved soil tilth.

Improved Cropping Systems

Recent reports have investigated the potential of organic agriculture to reduce greenhouse gas emissions. Organic systems of production increase soil organic matter levels through the use of composted animal manures and cover crops. Organic cropping systems also eliminate the emissions from the production and transportation of synthetic fertilisers. Components of organic agriculture could be implemented with other sustainable farming systems, such as conservation tillage, to further increase climate change mitigation potential.

Figure 1. Cover cropping between tree crops

Generally, conservation farming practices that conserve moisture, improve yield potential and reduce erosion and fuel costs also increase soil carbon. Examples of practices that reduce carbon dioxide emissions and increase soil carbon include direct seeding, field windbreaks, rotational grazing, perennial forage crops, reduced summer fallow and proper straw management. Using higher-yielding crops or varieties and maximising yield potential can also increase soil carbon.

Land Restoration

Land restoration and land use changes that encourage the conservation and improvement of soil, water and air quality typically reduce greenhouse gas emissions. Modifications to grazing practices, such as implementing sustainable stocking rates, rotational grazing and seasonal use of rangeland, can lead to greenhouse gas reductions. Converting marginal cropland to trees or grass maximises carbon storage on land that is less suitable for crops.

Irrigation and Water Management

Improvements in water use efficiency, through measures such as irrigation system mechanical improvements coupled with a reduction in operating hours; drip irrigation technologies; and center-pivot irrigation systems, can significantly reduce the amount of water and nitrogen applied to the cropping system. This reduces greenhouse emissions of nitrous oxide and water withdrawals.

Nitrogen Use Efficiency

Improving fertiliser efficiency through practices like precision farming using GPS tracking can reduce nitrous oxide emissions. Other strategies include the use of cover crops and manures; nitrogen-fixing crop rotations; composting and compost teas; and integrated pest management.

Methane Capture

Large emissions of methane and nitrous oxide are attributable to livestock waste treatment, especially in dairies. Agriculture methane collection and combustion systems include covered lagoons and complete mix and plug flow digesters. Anaerobic digestion converts animal waste to energy by capturing methane and preventing it from being released into the atmosphere. The captured methane can be used to fuel a variety of on-farm applications, as well as to generate electricity. Additional benefits include reducing odors from livestock manure and reducing labour costs associated with manure removal.

Biofuels

There is significant scientific controversy regarding whether biofuels — particularly those derived from oilseeds (biodiesel), feed corn (ethanol) or even from cellulosic sources — are carbon neutral. To ascertain the true climate neutrality of biofuels requires a careful life-cycle analysis of the specific biofuel under consideration. Also, an analysis is needed to understand what the global land use change implications will be if farmers grow more of a specific biofuel feedstock.

Soil Carbon: Benefits for Agriculture

Some argued that "creating farm and forestry systems with strong incentives for growing soil carbon could well be at the center of climate stabilisation." Thus, a new crop that farmers and ranchers may grow in the future is carbon. The Natural Resources Conservation Service, part of the USDA, has long been a promoter of managing carbon in efforts to improve soil quality.

As with any crop, farmers and ranchers need a market for this new crop, as well as a price that will make it more profitable to grow. From a broader social context, the questions of who will purchase this new crop and what is a fair price are also of private and public importance. Voluntary private carbon markets exist in the United States. Federal government markets are expected to be created soon. How to value carbon from the perspective of the individual farmer and rancher, as well as society at large, is the heart of understanding the role agriculture can play in carbon sequestration and climate stabilisation.

The two most frequently discussed systems to create value for offsetting greenhouse gas emissions are known as carbon taxation and cap and trade. Government subsidies are discussed less often, but will also play a role in greenhouse gas emission reductions.

CARBON TAX

By taxing every ton of carbon in fossil fuels or every ton of greenhouse gas companies emit, entities that emit greenhouse gases or use carbon-based fuels will have an incentive to switch to alternative renewable fuels, invest in technology changes to use carbonbased fuels more efficiently and in general adopt practices that would lower their level of greenhouse gas emissions. Thus a carbon or greenhouse gas emission tax values carbon in negative terms of tax avoidance. Those farms and ranches that emit or use less carbon-intensive fuels pay a smaller tax.

From the perspective of farmers and ranchers, a carbon tax would increase the direct and indirect costs of agricultural production. Farmers and ranchers use carbon-based fuels directly in the forms of petroleum and natural gas and indirectly in the forms of carbon-based fertilisers and pesticides and fuel-intensive inputs. Thus, a carbon tax could move farmers and ranchers to shift to systems of production that either eliminate the use of fossil fuels and inputs or at least improve the efficiency of their use.

However, proponents of carbon taxes have generally sought to exclude the agriculture sector from such taxation. For the most part, carbon tax proponents have been more interested in placing greenhouse gas emission taxes on upstream producers of the original source products. This includes coal, petroleum and natural gas producers and major emitters such as large electric utilities. Nonetheless, as people work to reduce greenhouse gas emissions, the potential to place a carbon tax on sectors like agriculture may become more likely.

Benefits of a Carbon Tax

A major benefit of a carbon or greenhouse gas emission tax would be the creation of a stream of tax revenue that the government could use to further induce the practice and technology changes necessary to lower greenhouse gas emissions.

For example, many of the current agriculture conservation programs, such as the Environmental Quality Incentive Program and the newer Conservation Stewardship Program, support improvements in soil quality and could be funded in part from emission or carbon taxes, thereby providing a revenue source to subsidise those who adopt or maintain emission-reduction practices or carbon sequestration activities. Tax revenues could also assist in the support of conservation programs like the Conservation Reserve Program, which works to keep sensitive and highly erodible lands out of production since these lands sequester soil carbon.

Another benefit of this approach is that a tax provides a clear and stable cost to current practices. A tax also makes it easier to determine changes that will be more profitable in a new cost environment. For instance, if a concentrated animal feeding operation understood the cost of their emissions as expressed by their emission tax, it would be easier for the operation to determine alternatives to current practices that would be cost efficient. At a high enough tax rate, installing methane digesters to lower greenhouse gas emission would become economically feasible. Finally, it has been argued that a carbon tax approach is cost effective in implementation, at least when compared to the cap-and-trade method of achieving greenhouse gas emissions reductions.

Downside of a Carbon Tax

Any tax results in discussions of where the burden of taxation lies and issues of equity. In short, taxation is about who pays and who does not. New taxes also often result in a public discussion of the fairness of the tax. There is logic to the argument that the burden of a carbon or greenhouse gas emission tax should be placed first and foremost on those who either create carbon-intensive fuels or those who are the largest emitters of greenhouse gases.

The greatest source of greenhouse gas emissions in the United States is the combustion of fossil fuels. Since agriculture uses a small percentage of fossil fuels, an argument can be made that the burden of taxation should not to fall on this sector. Still, agriculture is heavily dependent on fossil fuels and any carbon or greenhouse gas emission tax would likely be costly.

The ability of any individual farmer or rancher to pass on the increased costs of fossil fuels that this kind of taxation would create is much more limited than in other sectors of the economy. For instance, if a carbon tax is placed on diesel fuel, diesel fuel manufacturers can more easily pass on the tax burden to the consumers of the diesel.

The ability to pass on costs to consumers is greater in industries where there is little product substitution and where a few producers dominate the market. This is not the case for farmers and ranchers, given their relative lack of market concentration and power.

CAP-AND-TRADE SYSTEM

A government-sponsored cap-and-trade system would create a new market for greenhouse gas emissions by creating a new property right — the right to emit. The market is created by a government that sets a limit or cap on total greenhouse gas emissions allowed. Companies that emit greenhouse gases are issued emission permits that allow a certain amount of emissions. Companies and groups that exceed their allowed emissions must purchase offsets from other entities that pollute less than their allowance or from entities that sequester carbon.

These exchangeable emission permits, often called allowances, are measured in tons of carbon dioxide equivalents per year. Carbon dioxide equivalents provide a common measure for all greenhouse gas emissions and are calculated by converting greenhouse gases into carbon dioxide equivalents according to their global warming potential.

Over time, the government will continually lower the total level of allowances to meet an established level of acceptable total emissions. As the supply of allowances decreases, the value of the allowances will rise or fall depending on demand and on the ability of emitters to make necessary changes to reduce emissions or purchase offsets from groups more capable of reducing emissions.

Depending on the practices adopted, farmers and ranchers could be a source of inexpensive carbon reduction and capture the value of these allowances as offsets. In short, the value of offsets would become the market price of carbon equivalents. This would become the value of the new crop—carbon—that farmers and ranchers could grow.

For farmers and ranchers to provide carbon offsets for greenhouse gas emitters, farmers and ranchers must be willing to make longterm, or even permanent, changes in not only practices but perhaps whole systems of production. These changes also need to provide verifiable changes that result in true offsets of greenhouse gas emissions.

The issues of verifiability, permanence and what is known as *additionality* are critical to the success of agriculture's role in the cap-and trade system and the ultimate reduction of greenhouse gas emissions. Verifiability is critical because the system or practice change must result in a measurable change in the amount of carbon stored. The adoption of a no-till cultivation practice is thought to result in soil with higher carbon sequestration capacity. However, there is continuing scientific debate over whether the practice of continuous no-till does in fact lead to long-term additional storage of carbon in the soil.

The Chicago Climate Exchange (CCX) divided the United States into zones and allocated specific levels of carbon sequestration to each acre farmed in a particular zone under continuous no-till practices. While there may be some need to simplify the implementation of a nationwide soil carbon sequestration project related to tillage practice change, it is very doubtful that the actual carbon storage levels allocated can be achieved across areas that are so large.

Finally, the CCX does not verify the actual carbon storage as a result of the practice change, but only monitors that the practice is maintained during the life of the contract. Thus, it is doubtful the carbon offset truly matches actual carbon sequestered.

The issue of permanence is also critical. What happens after a farmer or rancher changes to a practice or system of production, is paid for carbon stored and then decides to change practices and potentially release the carbon that he or she was paid to sequester to offset emissions? Additionality refers to the issue that a farmer or rancher can only offer and be paid for an offset for a new sequestration of carbon, not for a practice or a system of production already in place. For instance, if a rancher developed a permanent wind shelter belt, that change in land use would likely result in new, or additional, carbon sequestration. However, a rancher who already developed a similar shelter belt would not be eligible for an offset because the rancher would not be providing additional carbon sequestration. Likewise, a farmer already engaged in conservation tillage would not provide additional carbon storage by maintaining that practice.

However, the current USDA Conservation Stewardship Program provides a possible payment structure that pays farmers to maintain practices. Additionality is also important because of the possibility that perverse incentives may be created that encourage farmers or ranchers to release carbon so that they can get paid to store it. A farmer practising no-till farming may decide to abandon the practice because of the new availability of per-acre payments and switch back to no-till at a later time. To address this and stop additional greenhouse gas emissions, the idea of offsets would need to be expanded to include farmers and ranchers already undertaking a practice or specific land use that stores soil carbon.

SUBSIDIES FOR AGRICULTURE

A final mechanism that could expand the ability of the agriculture sector to mitigate greenhouse gas emissions is one that is already well known — a direct subsidy. Many federal conservation programs provide incentives, known as cost shares, that help farmers and ranchers make changes in practices to conserve natural resources. Under the Conservation Stewardship Program and the Environmental Quality Incentive Program, farmers and ranchers can receive incentives to adopt new practices or receive support to maintain such practices. Though not designed to address climate change issues specifically, many federal conservation programs already provide public incentives to reduce greenhouse gas emissions.

In the future, conservation programs could be refocused to lower greenhouse emissions or increase carbon sequestration. Perhaps modifications of the Conservation Stewardship Program and the Environmental Quality Incentive Program could allow for longer contracts so that outcomes are reached and maintained. Also, the programs could add specific validation procedures to assure climate targets are met and sustained.

There is an immediate benefit to farmers and ranchers willing to make changes that meet the challenges of climate stabilisation. If sufficiently funded with outreach and technical assistance, efforts can be made to assure that all farmers and ranchers — regardless of their situation — take advantage of these programs. Finally, resources can be prioritised to different regions of the country or to specific practices or systems of production so programs can be cost-effective in reaching climate change goals.

Subsidies are a public cost, and this is a considerable downside. Furthermore, subsidies are based on the idea that the government can know and assure that the practices it pays for achieve the intended outcomes. For example, the federal government provides significant subsidisation of corn ethanol production.

Many argue that this changed the price of field corn and increased costs for people who use corn as animal feed and for other countries that import corn to feed people. There are also questions about how subsidies can reduce greenhouse gas emissions. Will subsidising a shift to a continuous no-till cultivation operation result in greater carbon sequestration? If the scientific understanding of the relationship between carbon sequestration and no-till is simply in error, then public dollars spent to change farmer behaviour would be wasted.

Paustian et al., estimated that it would take a price of at least $13 per ton of carbon dioxide equivalent ($50 per ton of carbon) per year to offset 70 million metric tons (MMT) of carbon dioxide equivalents. This would be a total public cost of close to $1 billion dollars per year for perhaps as long as 40 years. Also, this represents an offset of only 4 percent of total U.S. greenhouse gas emissions in 2004. Is this the least expensive way to reduce greenhouse gas emissions compared to alternative public expenditures? For instance, what if public dollars were committed to a research program to improve the gas mileage of automobiles? Recently, Sperow estimated an average cost to sequester carbon at $261 per ton of carbon. This is considerably higher than the Paustian estimate. While the difference between these studies can be explained by the fact that there is a wide regional variation in carbon sequestration capacity and how sequestration is accomplished, public costs would nonetheless be significant to achieve greenhouse gas emission reductions through subsidisation.

REFERENCES

Carrol, C. R., J.H. Vandermeer and P.M. Rosset. 1990. *Agroecology*. McGraw Hill Publishing Company, New York.

Jackson, Wes. 1985. *New Roots for Agriculture*. 2nd edition. University of Nebraska Press, Lincoln, NE. 150 p.

Pimentel, D., C. Harvey, P. Resosudarmo, et al. 1995. "Environmental and economic costs of soil erosion and conservation benefits". *Science*. Vol. 267, No. 245201. p. 1117-1123.

Sumner, D.R. 1982. "Crop rotation and plant productivity". In M. Recheigl, (ed). CRC *Handbook of Agricultural Productivity*, Vol. I CRC Press, Florida.

Thies, C. and T. Tscharntke. 1999. "Landscape structure, and biological control in agroecosystems". *Science* 285: 893-895.

Vandermeer, J. 1989. *The ecology of intercropping*. Cambridge University Press, Cambridge.

7

Cover Crops and Green Manures

A cover crop is any annual, biennial, or perennial plant grown as a monoculture (one crop type grown together) or polyculture (multiple crop types grown together), to improve any number of conditions associated with sustainable agriculture. Cover crops are fundamental, sustainable tools used to manage soil fertility, soil quality, water, weeds (unwanted plants that limit crop production potential), pests (unwanted animals, usually insects, that limit crop production potential), diseases, and diversity and wildlife, in agroecosystems. Agroecosystems are ecological systems managed by humans across a range of intensities to produce food, feed, or fiber. To a large degree, humans shape the ecological structure and function of natural processes that occur in agroecosystems. As agroecosystems often interact with neighboring natural ecosystems in agricultural landscapes, cover crops that improve the sustainability of agroecosystem attributes may also indirectly improve qualities of neighboring natural ecosystems. Farmers choose to grow specific cover crop types and to manage them in specific ways based on their own unique needs and goals. These needs and goals are influenced by biological, environmental, social, cultural, and economic factors of the food system within which farmers operate.

Cover crops also called "green manure" are used to manage a range of soil macronutrients and micronutrients. For example in Nigeria, the cover crop Mucuna pruriens (velvet bean) has been found to increase the availability of phosphorus in soil after a farmer applies rock phosphate. With respect to nutrients, the impact that cover crops have on nitrogen management has received by far the most attention by researchers and farmers, because nitrogen is often the most limiting nutrient in crop production. Cover crops known as "green manures" are grown and incorporated (by tillage) into the soil before reaching full maturity, and are intended to improve soil fertility and quality. They are commonly leguminous, meaning they are part of the fabaceae (pea) family. This family is unique in that all of the species in it set pods, such as bean, lentil, lupins and alfalfa. Leguminous cover crops are typically high in nitrogen and can often, to varying degrees, provide the required quantity of nitrogen for crop production that might normally be applied in chemical fertilizer form (called fertilizer replacement value). Another quality unique to leguminous cover crops is that they form symbiotic relationships with rhizobial bacteria that reside in legume root nodules. Lupins is nodulated by the soil microorganism Bradyrhizobium sp. (Lupinus). Bradyrhizobia are encountered as microsymbionts in other leguminous crops (Argyrolobium, Lotus, Ornithopus, Acacia, Lupinus) of Mediterranean origin. These bacteria convert biologically unavailable atmospheric nitrogen gas to biologically available mineral nitrogen through the process of biological nitrogen fixation.

TYPES OF COVER CROP

Winter Cover Crop

A winter cover crop is planted in late summer or fall to provide soil cover during the winter. Often a legume is chosen for the added benefit of nitrogen fixation. In northern states, the plant selected needs to possess enough cold tolerance to survive hard winters. Hairy vetch and rye are among the few selections that meet this need.

Winter cover crops include clovers, vetches, medics, and field peas. They are sometimes planted in a mix with winter cereal grains such as oats, rye, or wheat. Winter cover crops can be established by aerial seeding into maturing cash crops in the fall, as well as by drilling or broadcasting seed immediately following harvest.

Summer Green Manure Crop

A summer green manure occupies the land for a portion of the summer growing season. These warm-season cover crops can be used to fill a niche in crop rotations, to improve the conditions of poor soils, or to prepare land for a perennial crop. Legumes such as cowpeas, soybeans, annual sweetclover, sesbania, guar, crotalaria, or velvet beans may be grown as summer green manure crops to add nitrogen along with organic matter. Non-legumes such as sorghum-sudangrass, millet, forage sorghum, or buckwheat are grown to provide biomass, smother weeds, and improve soil tilth.

Living Mulch

A living mulch is a cover crop that is interplanted with an annual or perennial cash crop. Living mulches suppress weeds, reduce soil erosion, enhance soil fertility, and improve water infiltration. Examples of living mulches in annual cropping systems include overseeding hairy vetch into corn at the last cultivation, no-till planting of vegetables into subclover, sweetclover drilled into small grains, and annual ryegrass broadcast into vegetables. Living mulches in perennial cropping systems are simply the grasses or legumes planted in the alleyways between rows in orchards, vineyards, Christmas trees, berries, windbreaks, and field nursery trees to control erosion and provide traction.

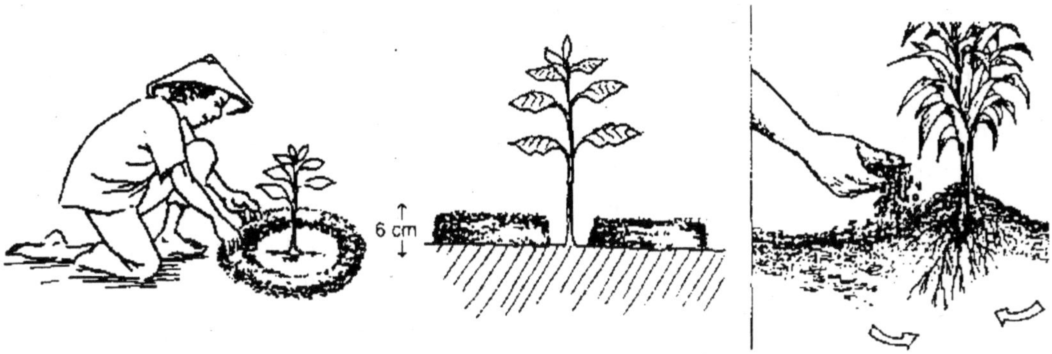

FIgure 1. Planting of mulch

Catch Crop

A catch crop is a cover crop established after harvesting the main crop and is used primarily to reduce nutrient leaching from the soil profile. For example, planting cereal rye following corn harvest helps to scavenge residual nitrogen, thus reducing the possibility of groundwater contamination. In this instance, the rye catch crop also functions as a winter cover crop. Short-term cover crops that fill a niche within a crop rotation are also commonly known as catch crops.

Forage Crop

Short-rotation forage crops function both as cover crops when they occupy land for pasturage or haying, and as green manures when they are eventually incorporated or killed for a no-till mulch. Examples include legume sods of alfalfa, sweet clover, trefoil, red clover, and white clover, as well as grass-legume sods like fescue-clover pastures. For maximum soil-improving benefits, the forage should not be grazed or cut for hay during its last growth period, to allow time for biomass to accumulate prior to killing.

BENEFITS OF COVER CROPS AND GREEN MANURES

Organic Matter and Soil Structure

A major benefit obtained from green manures is the addition of organic matter to the soil. During the breakdown of organic matter by microorganisms, compounds are formed that are resistant to decomposition—such as gums, waxes, and resins. These compounds—and the mycelia, mucus, and slime produced by the microorganisms—help bind together soil particles as granules, or aggregates. A well-aggregated soil tills easily, is well aerated, and has a high water infiltration rate. Increased levels of organic matter also influence soil humus. Humus—the substance that results as the end product of the decay of plant and animal materials in the soil—provides a wide range of benefits to crop production.

Sod-forming grass or grass-legume mixtures are important in crop rotations because they help replenish organic matter lost during annual cultivation. However, several years of sod production are sometimes required before measurable changes in humus levels occur. In comparison, annual green manures have a negligible effect on humus levels, because tillage and cultivation are conducted each year. They do replenish the supply of active, rapidly decomposing organic matter.

Table 1 shows dry matter production of several winter-annual legume cover crops grown in the southern US Approximately 2.2 tons per acre per year of crop residue is considered adequate to maintain soil organic matter at constant levels in continuously cropped soils.

Table 1. Average biomass yields and nitrogen yields of several legumes

Cover Crop	*Biomass tons/acre*	*Nitrogen lbs/acre*
Sweet Clover	1.75	120
Berseem Clover	1.10	70
Crimson Clover	1.40	100
Hairy Vetch	1.75	110

Nitrogen Production

Nitrogen production from legumes is a key benefit of growing cover crops and green manures. Nitrogen accumulations by leguminous cover crops range from 40 to 200 lbs. of nitrogen per acre. The amount of nitrogen available from legumes depends on the species of legume grown, the total biomass produced, and the percentage of nitrogen in the plant tissue. Cultural and environmental conditions that limit legume growth—such as a delayed planting date, poor stand establishment, and drought—will reduce the amount of nitrogen produced. Conditions that encourage good nitrogen production include getting a good stand, optimum soil nutrient levels and soil pH, good nodulation, and adequate soil moisture.

The portion of green-manure nitrogen available to a following crop is usually about 40% to 60% of the total amount contained in the legume. A hairy vetch crop that accumulated 180 lbs. N per acre prior to plowing down will contribute approximately 90 lbs. N per acre to the succeeding grain or vegetable crop.

To determine how much nitrogen is contained in a cover crop, an estimate is needed of the yield of above-ground herbage and its percentage of nitrogen. The procedure involves taking a field sample, drying it, weighing it, and sending a sample off for forage analysis, which includes an estimate of protein content. Once the protein content is known, simply divide it by 6.25 to obtain the percentage of nitrogen contained in the cover crop tissue. Finally, to obtain pounds of legume nitrogen per acre.

Forage legumes are valuable in rotations because they generate income from grazing or haying and still contribute nitrogen from regrowth and root residues. A high percentage of biologically fixed nitrogen is in the top growth. (Table 2)

Table 2. Percent nitrogen in legume tops and roots.

Crop	*Tops*	*Roots*
	%N	*%N*
Soybeans	93	7
Vetch	89	11
Cowpeas	84	16
Red Clover	68	32
Alfalfa	58	42

A rapid increase in soil microorganisms occurs after a young, relatively lush green manure crop is incorporated into the soil. The soil microbes multiply to attack the freshly incorporated plant material. During microbial breakdown, nutrients held within the plant tissues are released and made available to the following crop.

Soil Microbial Activity

Factors that influence the ability of microorganisms to break down organic matter include soil temperature, soil moisture, and carbon to nitrogen (C:N) ratio of the plant material. The C:N ratio of plant tissue reflects the kind and age of the plants from which it was derived. (Table 3) As plants mature, fibrous (carbon) plant material increases and protein (nitrogen) content decreases. The optimum C:N ratio for rapid decomposition of organic matter is between 15:1 and 25:1.

C:N ratios above 25:1 can result in nitrogen being "tied up" by soil microbes in the breakdown of carbon-rich crop residues, thus pulling nitrogen away from crop plants. Adding some nitrogen fertiliser to aid the decomposition process may be advisable with these high carbon residues. The lower the C:N ratio, the more N will be released into the soil for immediate crop use.

Table 3. Common C:N ratios of cover crops.

Organic Material	*C:N Ratio*
Young rye plants	14:1
Rye at flowering	20:1
Hairy vetch	10:1 to 15:1
Crimson clover	15:1
Corn stalks	60:1
Sawdust	250:1

The C:N ratio is more a function of the plant's N content than its carbon content. Most plant materials contain close to 40% carbon. To determine the C:N ratio of any plant material, divide 40% by its nitrogen content. For example let's say hairy vetch contains 4.2% nitrogen: 40/4.2= a C:N ratio of 9.5. A procedure for determining the nitrogen content of cover crop biomass was previously addressed in the section on nitrogen production. Estimating the nitrogen contribution of a cover crop is very helpful when adjusting N-fertiliser rates to account for legume nitrogen. Table 3 provides a nice comparison of the typical C:N ratios that can be found in different types of crop residues. The important point is that lush green manures are richer in nitrogen relative to carbon, especially in comparison to highly lignified crop residues like corn stalks. It will take a lot longer for soil microbes to break down corn stalks than fresh hairy vetch.

Nutrient Enhancement

In addition to nitrogen from legumes, cover crops help recycle other nutrients on the farm. Nitrogen (N), phosphorous (P), potassium (K), calcium (Ca), magnesium (Mg), sulfur (S), and other nutrients are accumulated by cover crops during a growing season. When the green manure is incorporated, or laid down as no-till mulch, these plant-essential nutrients become slowly available during decomposition. Dr. Greg Hoyt developed a method for estimating nutrient accruement by cover crops in order to reduce the soil test recommendation of fertiliser for the following crop. Table 4 shows the biomass and nutrients accumulated by several cover crops he worked with.

Certain broad-leaved plants are noted for their ability to accumulate minerals at high concentrations in their tissue. For example, buckwheat, lupine, and sweetclover are noted for their ability to extract P from soils. Likewise, alfalfa and other deep-rooting green manures scavenge nutrients from the subsoil and translocate them upwards to the surface rooting zone, where they become available to the following crop. The breakdown of green manures in soil influences mineral nutrient availability in another way.

Table 4. Biomass yield and nutrient accruement by selected cover crops.

Crop	Biomass lbs/ac	Nitrogen lbs/ac	Potassium lbs/ac	Phosphorus lbs/ac	Magnesium lbs/ac	Calcium lbs/ac
Hairy Vetch	3,260	141	133	18	18	52
Crimson Clover	4,243	115	143	16	11	62
Austrian Winter Peas	4,114	144	159	19	13	45
Rye	5,608	89	108	17	8	22

During decomposition of organic matter, carbonic and other organic acids are formed as a byproduct of microbial activity. These organic acids react with insoluble mineral rocks and phosphate precipitates, releasing phosphates and exchangeable nutrients.

Rooting Action

The extensive root systems of some cover crops are highly effective in loosening and aerating the soil. In Australian wheat experiments, the taproots of a blue lupine cover crop performed like a "biological plow" in penetrating compacted soils. When cover crops are planted after a subsoiling treatment, they help extend the soil-loosening effects of the deep tillage treatment. The rooting depths of several green manures grown under average conditions are listed in (Table 5).

Table 5. Typical rooting depths of several green manure crops.

Depth (feet)	Green Manure Crop
5 to 7	Red Clover, Lupine, Radish, Turnips
3 to 5	Common Vetch, Mustard, Black Medic, Rape
1 to 3	White Clover, Hairy Vetch

Weed Suppression

Weeds flourish on bare soil. Cover crops take up space and light, thereby shading the soil and reducing the opportunity for weeds to establish themselves. The soil-loosening effect of deep-rooting green manures also reduces weed populations that thrive in compacted soils. The primary purpose of a non-legume green manure—such as rye, millet, or sudangrass—is to provide weed control, add organic matter, and improve soil tilth. They do not produce nitrogen. Thus, whenever possible, annual grain or vegetable crops should follow a legume green manure to derive the benefit of farm-produced nitrogen.

Providing weed suppression through the use of allelopathic cover crops and living mulches has become an important method of weed control in sustainable agriculture. Allelopathic plants are those that inhibit or slow the growth of other nearby plants by releasing natural toxins, or "allelochemicals." Cover crop plants that exhibit allelopathy include the small grains like rye and summer annual forages related to sorghum and sudangrass.

The mulch that results from mowing or chemically killing allelopathic cover crops can provide significant weed control in no-till cropping systems. Living mulches suppress weeds during the growing season by competing with them for light, moisture, and nutrients.

Soil and Water Conservation

When cover crops are planted solely for soil conservation, they should provide a high percentage of ground coverage as quickly as possible. Most grassy and non-legume cover crops, like buckwheat and rye, fulfill this need well. Of the winter legumes, hairy vetch provides the least autumn ground cover because it puts on most of its above-ground growth in the spring.

Consequently, it offers little ground cover during the erosion-prone fall and winter period. Sowing a mix of leguminous and grassy-type cover crops will increase the ground coverage, as well as provide some nitrogen to the following crop. The soil conservation benefits provided by a cover crop extend beyond protection of bare soil during non-crop periods.

The mulch that results from a chemically or mechanically killed cover crop in no-till plantings increases water infiltration and reduces water evaporation from the soil surface. Soil cover reduces soil crusting and subsequent surface water runoff during rainy periods. Retention of soil moisture under cover crop mulches can be a significant advantage.

VEGETATION MANAGEMENT TO CREATE A COVER CROP MULCH

Herbicides are the most commonly used tools for cover crop suppression in conservation tillage systems. Non-chemical methods include propane flamers, mowing and mechanical tillage. Mowing a rye cover crop when it heads out in late spring provides sufficient kill. The rye must be in the pollination phase, or later, to be successfully killed. When the anthers are fully extended and you can thump the stalk and pollen falls down, it is time to mow. If mowed earlier, it just grows back. Flail mowers generally produce more uniformly distributed mulch than do rotary cutters, which tend to windrow the mulch to one side of the mower.

Sickle bar mowers create fairly uniform mulch, but the unchopped rye stalks can be more difficult to plant into. If late spring weather continues cool and wet, more rye regrowth will occur than if the weather remains warm and drier. Typically, if rye is mowed at the pollination stage, regrowth is minimal and not a problem to crops grown in the mowed mulch. Flail mowing, or rolling with rolling disk colters spaced at 4 inches, was usually as effective as herbicides in killing hairy vetch, crimson clover and subterranean clover.

Timing is a key factor when using mowing or rolling to control cover crops. Mechanical control was most effective when the legumes were in the seed formation growth phase or when stem lengths along the ground exceeded 10 inches. If mowing was followed with a pre-plant herbicide application of Atrazine, the legume kill was even more effective.

The blades are pitched to 15 degrees, allowing the blades to penetrate the soil and provide a slight lifting action. Mounted on the same toolbar behind the cutter blades is a rolling basket to flatten and distribute the undercut cover crop. The undercutter was tried on several cover crops and effectively killed crimson clover, hairy vetch, rye, and barley. These undercutters could be made from locally available stock by innovative tinkerers.

Planting cover crops known to readily winter-kill is another non-chemical means of vegetation management. Spring oats, buckwheat, and sorghum fill this need. They are fall-planted early enough to accumulate some top growth before freezing temperatures kill them. In some locations, oats will not be completely killed and some plants will regrow in the spring. Winter-killed cover crops provide a dead mulch through the winter months instead of green cover. They are used primarily in regions where precipitation is limited, such as West Texas, and in situations where farmers want to plant early in the spring and avoid overwintered cover crops altogether.

LIMITATIONS OF COVER CROPS

The recognised benefits of green manuring and cover cropping—soil cover, improved soil structure, nitrogen from legumes—need to be evaluated in terms of cash returns to the farm as well as the long-term value of sustained soil health. For the immediate growing season, seed and establishment costs need to be weighed against reduced nitrogen fertiliser requirements and the effect on cash crop yields. Water consumption by green manure crops is a concern and is pronounced in areas with less than 30 inches of precipitation per year. Still, even in the fallow regions of the Great Plains and Pacific Northwest, several native and adapted legumes seem to have potential for replacing cultivation or herbicides in summer fallow.

There is always additional management required when cover crops of any sort are added to a rotation. Turning green manures under or suppressing cover crops requires additional time and expense, compared to having no cover crop at all. Insect communities associated with cover crops work to the farmer's advantage in some crops and create a disadvantage in others.

Certain living mulches enhance the biological control of insect pests of summer vegetable crops and pecan orchards by providing favorable habitats for beneficial insects. On the negative side, winter legumes that harbor catfacing insects such as the tarnished plant bug, stink bug, and plum curculio can pose problems for apple or peach orchardists. Nematodes encouraged by certain legumes on sandy soils are another concern of farmers, as are cutworms in rotations following grain or grass crops.

COVER CROPS IN ROTATION

Cover crops can fit well into many different cropping systems during periods of the year when no cash crop is being grown. Even the simplest corn/soybean rotation can accommodate a rye cover crop following corn, which will scavenge residual nitrogen and provide ground cover in the fall and winter. When spring-killed as a no-till mulch, the rye provides a water-conserving mulch and suppresses early-season weeds for the following soybean crop.

Hairy vetch can be planted behind soybeans to provide nitrogen for corn the following spring. Hairy vetch is *not* a good cover crop to use when small grains are included in the rotation—if the vetch ever goes to seed it can become a terrible weed in the small grain crop. In these cases, crimson clover, sweet clover, or red clover should be used, depending on location. Many vegetable rotations can accommodate cover crops as well. Buckwheat can follow lettuce and still be tilled down in time for fall broccoli. Hairy vetch works well with tomatoes and other warm-season vegetables. The vetch can be killed by flail mowing and tomato sets planted into the mulch.

PEST MANAGEMENT BENEFITS OF COVER CROPS

In addition to the soil improving benefits, cover crops can also enhance many pest management programs. Ecologists tell us that stable natural systems are typically diverse, containing many different types of plants, arthropods, mammals, birds, and microorganisms. Growing cover crops adds diversity to a cropping system. In stable systems, serious pest outbreaks are rare, because natural controls exist to automatically bring populations back into balance.

Farmers and researchers in several locations have observed and documented increased beneficial insect numbers associated with cover crops. The cover crops provide pollen, nectar, and a physical location for beneficial insects to live while they search for pest insects. Conservation tillage proves a better option than tilling because it leaves more crop residue on the surface to harbor the beneficial insects. Strip tilling or no-tillage disturbs a minimum of the existing cover crop that harbors beneficial insects. Cover crops left on the surface may be living or in the process of dying. At either of these stages they protect beneficials. Once the main crop is growing, the beneficials move onto it. By having the cover crop in place early in the growing season, the population of beneficials is much higher sooner in the growing season than would be the case if only the main crop were serving as habitat for the beneficials. Innovative farmers are paving the way by interplanting cover crops with the main crop and realising pest management benefits as a result.

Ground preparation for cotton planting is begun about 10 days prior to planting by tilling 14-inch wide strips into the lupines. Herbicides are applied to the strips at that time and row middles remain untouched. The remaining lupines provide beneficial insect habitat and also serve as a smother crop to curtail weeds and grasses. The lupines in the row middles can be tilled in later in the season to release more legume nitrogen. Planting cotton into strip-killed crimson clover improves soil health, cuts tillage costs, and allows him to grow cotton without any insecticides and only 30 pounds of nitrogen fertiliser.

ECONOMICS OF COVER CROPS

The most obvious direct economic benefit derived from legume cover crops is nitrogen fertiliser savings. In most cases these savings can offset cover crop establishment costs. Indirect benefits include herbicide reduction in the case of an allelopathic rye cover crop, reduction in insect and nematode control costs in some cases, protection of ground water by scavenging residual nitrate, and water conservation derived from a no-till mulch. Longer-term benefits are derived from the buildup of organic matter resulting in increased soil health. Healthy soils cycle nutrients better, don't erode, quickly absorb water after each rain, and produce healthy crops and bountiful yields.

With annual cover crops, the highest cost is seed. Hairy vetch and crimson clover typically range from 50¢ to $1.50 per pound. With a 20-pound per acre seeding rate, seed costs range from $10 to $30 per acre. With a 25-pound seeding rate at 85¢/lb and a $6.50 no-till drilling cost, it would cost $28 to plant an acre of this cover crop. Hairy vetch was compared to a winter wheat cover crop or no cover crop at two different locations (coastal plain and piedmont). Corn was grown following the cover crops. Nitrogen fertiliser was used with the cover crops at varying rates. The most profitable cover crop and nitrogen fertiliser combination used more than 100 lbs of additional nitrogen per acre plus the cover crop. At $2.50 per bushel corn price, highest returns at the coastal plain location were realised with 120 lbs of additional nitrogen per acre. Profits were as follows: $53.75 per acre from no cover crop, $95.62 from hairy vetch, and $32.47 from winter wheat cover crop. All corn crops needed additional nitrogen. Lower N rates were less profitable. At the piedmont location, also with $2.50 corn, winter fallow was most profitable at $68.03 with 40 lbs per acre additional N, hairy vetch was profitable at $56.57 with 40 lbs per acre, and winter wheat was profitable at $30.12 with 100 lbs of additional nitrogen.

Optimum nitrogen rates for corn were determined when corn followed four cover crops, compared to a winter fallow treatment. Corn was grown following each cover crop treatment at various nitrogen rates over a three-year period. The results are shown in (Table 6). The optimum nitrogen rate is the rate above which no additional yield increases are realised. The researchers concluded that cover crops can benefit a succeeding corn crop not only by supplying nitrogen but also by increasing maximum yield of the system.

Table 6. Optimum Nitrogen rates and profitability of several cover crops.

Cover Crop	*Corn Yield bu/acre*	*Optimum N Rate lbs N/acre*
No cover crop	142	100
Winter wheat	142	126
Hairy vetch	148	79
Austrian winter peas	153	107
Crimson clover	148	94

The economic value of these nitrogen replacements can be calculated by using a local nitrogen price. These costs can then be compared to cover crop seed and planting costs. These simple nitrogen cost comparisons do not take into account the benefits of improved soil tilth and increased water infiltration resulting from cover crops.

Economic returns above direct expenses for no-till corn were $64 greater with hairy vetch plus 90 lbs of nitrogen fertiliser per acre than with no cover crop plus the same nitrogen rate. This advantage was mostly due to the yield increase under the legume cover crop of 36 bushels per acre. Some researchers have stated that advantages of legume cover crops can only be realised if they increase yields of a following crop over yields obtained from no cover crop. In other words, the nitrogen replacement value is insufficient to offset the establishment costs of the cover crop without an increase in crop yield. When these yield increases beyond the nitrogen benefit occur, they are due to improved soil water use efficiency and other soil health benefits from the cover crop.

MANURES FOR ORGANIC CROP PRODUCTION

Livestock manure is traditionally a key fertiliser in organic and sustainable soil management. It is most effectively used in combination with other sustainable practices. These include crop rotation, cover cropping, green manuring, liming, and the addition of other natural or biologically friendly fertilisers and amendments. In organic production, manure is commonly applied to the field in either a raw or composted state.

There are clear restrictions on the use of raw manure in organic farming. These restrictions are detailed in the National Organic Program (NOP) Regulations, which constitute the federal standard for organic production.

Raw Manure Use

Raw manure is an excellent resource for organic crop production. It supplies nutrients and organic matter, stimulating the biological processes in the soil that help to build fertility. Still, a number of cautions and restrictions are in order, based on concerns about produce quality, food contamination, soil fertility imbalances, weed problems, and pollution hazards.

Contamination

Some manures may contain contaminants such as residual hormones, antibiotics, pesticides, disease organisms, and other undesirable substances. Since many of these can be eliminated through high-temperature aerobic composting, this practice is recommended where low levels of organic contaminants may be present. Caution is advised, however, as research has demonstrated that *Salmonella and E. coli* bacteria appear to survive the composting process much better than previously thought. The possibility of transmitting human diseases discourages the use of fresh manures (and even some composts) as preplant or sidedress fertilisers on vegetable crops—especially crops that are commonly eaten raw.

In February 2000, the issue of manure use on organic farms was highlighted on the television news program *20/20*. The segment suggested that fertilisation with livestock manures made organic foods more dangerous than other food products in the marketplace. The show's producers arranged for a sampling of various organic and nonorganic vegetables from store shelves and tested for the presence of *E. coli*. The samples of both organic and nonorganic produce were generally free of serious contamination. The exceptions were bagged sprouts and mesclun salad mix. Of these, more *E. coli* contamination was observed on the organic samples. It was largely on this basis that the news program challenged organic farming.

The attack was embarrassing to the organic industry and forced its membership to undertake a lot of "damage control," despite the fact that the allegations were contrived and based on poor science. The sampling was not statistically significant. The show failed to point out that the specific test used does not distinguish between pathogenic and benign forms of *E. coli*. Also ignored was the obvious fact that conventional farmers use manure, too! Furthermore, the reporter failed to disclose the vested interests of the individual bringing the charges, presenting him instead as a former official with the Agriculture Department. John Stossel—the journalist responsible for the *20/20* report—subsequently issued an apology and a correction.

Unlike conventional farmers, who have only safety guidelines regarding manure use, certified organic farmers must follow stringent protocols. Raw manure may not be applied to food crops within 120 days of harvest where edible portions have soil contact (i.e., most vegetables, strawberries, etc.); it may not be applied to food crops within 90 days of harvest where edible portions do not have soil contact (i.e., grain crops, most tree fruits). Such restrictions do not apply to feed and fiber crops.

Organic substances are not the only contaminants found in livestock manures. Heavy metals can be a problem, especially where industrial-scale production systems are used. Concerns over heavy-metal and other chemical contamination have dogged the use of poultry litter as an organically acceptable fertiliser in Arkansas, where it's readily and cheaply available. Heavy-metal contamination is also a concern with composted sewage sludge (biosolids)—a major reason for its being prohibited from certified organic production. Under federal organic standards, certifiers may require testing of manure or compost if there is reason to suspect high levels of contamination.

Produce Quality Concerns

It has long been acknowledged that improper use of raw manure can adversely affect the quality of such vegetable crops as potatoes, cucumbers, squash, turnips, cauliflower, cabbage, broccoli, and kale. As it breaks down in the soil, manure releases chemical compounds such as skatole, indole, and other phenols. When absorbed by the growing plants, these compounds can impart off-flavors and odors to the vegetables. For this reason, raw manure should not be directly applied to vegetable crops.

Fertility Imbalances

Raw manure use has often been associated with imbalances in soil fertility. There are several causal factors:

- Manure is often rich in specific nutrients like phosphate or potash. While these nutrients are of great benefit to crops, repeated applications of manure can result in their building to detrimental levels. A good example is the overuse of broiler litter in the mid-South, which has put excessive phosphate in the soil and polluted surface waters. Nutrient excesses also "tie up" other minerals. Excessive phosphate interferes with plant uptake of both copper and zinc; excessive potash can restrict boron, manganese, and even magnesium.
- Continual manure use tends to acidify soil. As manure breaks down it releases various organic acids that assist in making soil minerals available—a benefit of manure that is often unrecognised. Over time, however, this process depletes the soil of calcium and causes pH levels to fall below the optimum for most crops. Manures do supply some calcium, but not enough to counterbalance the tendency toward increased acidity. Possible exceptions include caged-layer manure and manure from dairy operations where barn lime is used.
- When fresh manure containing large amounts of nitrogen and salts is applied to a crop, it can have the same effects as excessive applications of soluble commercial fertilisers—it can burn seedling roots, reduce immunity to pests, and shorten produce shelf life. Excessive salinity is often associated with heavy applications of feedlot manure in regions where little leaching naturally occurs—as in most western states.

Table 7: Approximate NPK Values of Various Animal Manures

Animal	*% nitrogen*	*% phosphoric acid*	*% potash*
Dairy cow	0.57	0.23	0.62
Beef steer	0.73	0.48	0.55
Horse	0.70	0.25	0.77
Swine	0.49	0.34	0.47
Sheep/Goat	1.44	0.50	1.21
Rabbit	2.40	1.40	0.60
Chicken	1.00	0.80	0.39

To avoid manure-induced imbalances, continually monitor soil fertility, using appropriate soil tests. Then apply lime or other supplementary fertilisers and amendments to ensure soil balance, or restrict application levels if needed. Understanding the soil's needs is only part of the equation. You must also know the nutrient content of the manure you're applying. Standard fertiliser values should be used only for crude approximations. The precise nutrient content of any manure is dependent not only on livestock species, but also on the ration fed, the kind of bedding used, amount of liquid added, and the kind of capture and handling system employed.

Some traditional assumptions about manure composition may need to be updated. Because of the abundance of sulfur in rations, manure has long been recognised as a good source of sulfur. However, less sulfur is applied to crops in contemporary high-analysis fertilisers, and atmospheric deposition has been decreased by pollution controls. Sulfur deficiencies are appearing in many soils, and levels in manure may also be diminished. It is advisable to test manure as you would test the soil, in order to assess its fertiliser value.

Weed Problems

Use of raw manures has often been associated with increased weed problems. Some manure contains weed seed, often from bedding materials like small-grain straw and old hay. High-temperature aerobic composting can greatly reduce the number of viable weed seeds. In many cases, however, the lush growth of weeds that follows manuring does not result from weed seeds in the manure, but from the stimulating effect manure has on seeds already present in the soil.

The flush of weeds may result from enhanced biological activity, the presence of organic acids, an excess of nitrates, or some other change in the fertility status of the soil. Depending on the weed species that emerge, the problem may be related to the sort of fertility imbalances described above. Excesses of potash and nitrogen in particular can encourage weeds. Monitor the nutrient contents of soil and manure and spread manure evenly to reduce the incidence of weed problems.

Pollution

When the nutrients in raw or composted manure are eroded or leached from farm fields or holding areas, they present a potential pollution problem, in addition to being a resource lost to the farmer. Leached into groundwater, nitrates from manure and fertilisers have been linked to a number of human health problems. Flushed into surface waters, nutrients can cause eutrophication of ponds, lakes, and streams. Excess nitrates from farms and feedlots in the Mississippi Basin are deemed the primary cause of the Gulf of Mexico Dead Zone—a hypoxic (oxygen deprived) area about the size of New Jersey that now threatens the shrimp, fishing, and recreational industries off the Louisiana and Texas coasts.

The manner in which manure is collected and stored prior to field use affects the stabilisation and conservation of valuable nutrients and organic matter. Reducing manure run-off and leaching losses from fields is a matter of both volume and timing.

Manure application far in excess of crop needs greatly increases the chances of nutrient loss, especially in high-rainfall areas. Manure spread on bare, frozen or erodible ground is subject to run-off, especially where heavy winter rains are common. Under some conditions, however, winter-applied manure can actually slow run-off and erosion losses from fields, likely by acting as a light organic mulch.

Sheet-composting manure or applying it to growing cover crops are two advisable means of conserving manure nutrients. Grass cover crops, such as rye and ryegrass, are especially good as "catch crops"—cover crops grown to absorb soluble nutrients from the soil profile to prevent them from leaching. It is a sound strategy, therefore, to apply manure to growing catch crops or just prior to planting them. Note that both sheet composting and applying to cover crops have trade-offs. Sheet composting improves the capture of ammonia nitrogen from manure, but requires tillage, which leaves the soil bare and vulnerable to erosion and leaching losses. Surface-applying to cover crops eliminates most leaching and erosion losses but increases ammonia losses to the atmosphere.

Composted Manures

An effective composting process converts animal wastes, bedding, and other raw products into humus—the relatively stable, nutrient-rich, and chemically active organic fraction found in fertile soil. In stable humus, there is practically no free ammonia or soluble nitrate, but a large amount of nitrogen is tied up as proteins, amino acids, and other biological components. Other nutrients are stabilised in compost as well. Composting livestock manure reduces many of the drawbacks associated with raw manure use. Good compost is a "safe" fertiliser.

Low in soluble salts, it doesn't "burn" plants. It's also less likely to cause nutrient imbalances. It can safely be applied directly to growing vegetable crops. Many commercially available organic fertilisers are based on composted animal manures supplemented with rock powders, plant by-products like alfalfa meal, and additional animal by-products like blood, bone, and feather meals. The quality of compost depends on the feedstuffs used to make it. Unless it is supplemented in some way, composted broiler litter—though more stable than raw litter—will be abundant in phosphates and low in calcium.

Continued applications may lead to imbalanced soil conditions in the long term, as with some raw manures. Soil and compost testing to monitor nutrient levels is strongly advised. While composting can degrade many organic contaminants, it cannot eliminate heavy metals. In fact, composting concentrates metals, making the contaminated compost, pound for pound, more potentially hazardous than the manure it was created from.

Broiler litter and broiler litter composts have been restricted from certified organic production in the mid-South largely for this reason. Arsenic—once used in chicken feed as an appetite stimulant and antibiotic—was a particular concern. Since the precise composition of commercial livestock feeds is proprietary information, arsenic may still be an additive in formulations in some regions.

A more recent concern is the inclusion of additional copper in poultry diets and its accumulation in the excreted manure. While copper is an essential plant nutrient, an excessive level in the soil is toxic. This concern is most relevant to organic horticultural producers, who often apply significant amounts of copper as fungicides and bactericides, increasing the hazard of buildup in the soil. Whenever you import large amounts of either composted or raw manure onto the farm, it is wise to inquire about the feeding practices at the source or have the material tested.

REFERENCES

Allison, F.E. 1973. *Soil Organic Matter and Its Role in Crop Production.* Elsevier Scientific Pub. Co. Amsterdam. 637 p.

Schmid, O., and R. Klay. 1984. *Green Manuring: Principles and Practice.* Woods End Agricultural Institute, Mt. Vernon, Maine. Translated by W. F. Brinton, Jr., from a publication of the Research Institute for Biological Husbandry. Switzerland. 50 p.

Rasmussen, R.R. et al. 1980. "Crop residue influences on soil carbon and nitrogen in a wheat-fallow system". Soil Science Society of America Proceedings. Volume 44. p. 596-600.

Sarrantonio, Marianne. 1994. *Northeast Cover Crop Handbook.* Rodale Institute, Emmaus, PA. 118 p.

Piper, C.V., and A.J. Pieters. 1922. *Green Manuring.* USDA Farmer's Bulletin 1250. 45 p.

McLeod, E. 1982. *Feed the Soil.* Organic Agriculture Research Institute, Graton, CA. 209 p.

May, J.T. 1981. "Organic matter in nursery soils". p. 52-59. In: Proceedings of the 1981 Southern Nursery Conference.

8

Biological Pest Control

Biological control of pests in agriculture is a method of controlling pests that relies on predation, parasitism, herbivory, or other natural mechanisms. It can be an important component of integrated pest management (IPM) programs.

'Biological pest control' is defined as the reduction of pest populations by natural enemies and typically involves an active human role. Natural enemies of insect pests, also known as biological control agents, include predators, parasitoids, and pathogens. Biological control agents of plant diseases are most often referred to as antagonists. Biological control agents of weeds include herbivores and plant pathogens. Predators, such as lady beetles and lacewings, are mainly free-living species that consume a large number of prey during their lifetime.

Parasitoids are species whose immature stage develops on or within a single insect host, ultimately killing the host. Most have a very narrow host range. Many species of wasps and some flies are parasitoids. Pathogens are disease-causing organisms including bacteria, fungi, and viruses. They kill or debilitate their host and are relatively specific to certain insect groups. There are three basic types of biological control strategies; conservation, classical biological control, and augmentation.

The conservation of natural enemies is probably the most important and readily available biological control practice available to homeowners and gardeners. Natural enemies occur in all areas, from the backyard garden to the commercial field. They are adapted to the local environment and to the target pest, and their conservation is generally simple and cost-effective.

Lacewings, lady beetles, hover fly larvae, and parasitised aphid mummies are almost always present in aphid colonies. Fungus-infected adult flies are often common following periods of high humidity. These naturally occurring biological controls are often susceptible to the same pesticides used to target their hosts. Preventing the accidental eradication of natural enemies is termed simple conservation.

BIOLOGICAL CONTROL IN A BIODIVERSITY CONTEXT

There are many organisms in agro-ecosystems which may have a value, negative or positive, to crop production. Natural enemies comprise a subset, the "borders" of which is sometimes difficult to define. Species in agro-ecosystems interact in food chains. Predators, parasitoids and pathogens attack insect crop pests, while plant feeding insects and pathogens attack weeds of crops. These natural enemies have a clear and direct biological control function. However, other organisms may have a biological control effect without actually killing pests directly, because of the web-like nature of relations between species in crops. For instance, crop protection may be conferred by:

- herbivores which preferentially feed on weeds which compete with crops, thereby improving crop vigour and production
- herbivores on crop or wild plants which support populations of natural enemies which also attack crop pests
- benign micro-organisms which colonise crops and their habitats, thereby reducing colonisation by pathogens
- plants or invertebrates which produce chemicals repellents to pest species
- non-crop plants which are attractive to pests and thereby reduce their numbers on crops

The sheer diversity of interacting invertebrates, micro-organisms and plants in agro-ecosystems makes these indirect biological control effects likely. For instance a single rice crop in Asia may have almost 600 species of higher animals and plants interacting in food webs not to mention the diversity of micro-organisms associated with these species and with soils. It is important, therefore, to note that those pest-destroying natural enemies which I will consider here represent only a fraction of the biodiversity which is important to biological control in agroecosystems.

Other, beneficial crop associated biodiversity, including pollinators and species which contribute to soil fertility, are, like natural enemies, often insects, fungi or other micro-organislsm, and occur in similar habitats. Hence they may be subject to similar agricultural processes. These include threats, such as pesticide use against beneficial arthropods, as well as benefits. Because beneficial, biological control effects are generated directly or indirectly by a wide range of crop-associated organisms, not just natural enemies, and because other beneficial organisms in crops are affected by agricultural practices in a similar way to natural enemies, it is neither easy, nor perhaps desirable, to separate biological control organisms out for separate consideration of biodiversity conservation and use.

Natural Enemy Diversity and Function

From the perspective of genetic resource conservation and use, the most important feature of natural enemies is their remarkable diversity. It has been suggested that as much as 50% of extant species are natural enemies in the categories above. The Millennium Ecosystem Assessment estimates the total number of species on earth to exceed ten million, based on the work of Groombridge and Jenkins. About 8m of these are insects and myriapods, 1.5m fungi, 0,75m chelicerates, 0.65m protoctists and 0.4m nematodes. Natural enemies constitute a substantial proportion of insects, fungi and chelicerate species, and are well represented in protoctista and 0.4m nematodes. Hence, there are probably millions of natural enemy species. For the three largest taxa, insects, myriapods, fungi and chelicerates, the analysis suggests that, of their approximately 10.25m estimated species, about 9 million (88%) are yet to be described.

By contrast, all crop species are described and thousands of specific plant genetic resource accessions and varieties have been characterised in the last decades. Even if we consider plant species generally, as a future resource for new crops, recent estimates suggest there are at most a few hundred thousand species, the great majority of which are named. Hence, living resources for biological control are far more diverse and far less well known than plant resources for agriculture.

In order to understand natural enemies in agro-ecosystems so that we might best conserve and use them, it is also helpful to appreciate aspects of their *specificity* and *functionality*. Natural enemy species may be highly specific, that is, they may attack and reproduce on only a single pest species. This is one explanation for the enormous diversity of natural enemies: because of their high specificity, natural enemies are *at least as* diverse as their prey or hosts. When you consider that these prey and hosts comprise insects, plants and fungi, themselves enormously diverse, it is clear how such natural enemy diversity can arise. However, other natural enemy species may attack a range of pest species, and we call these generalists.

Within a particular natural enemy taxon, both specialist and generalist species may be found. For instance, for insect parasitoids, 60% of the 214 species of parasitoids in the family Braconidae attacking plant-feeding flies in the family Agromyzidae in Britain are specific to only one species, and of 514 species of parasitoids in the family Ichneumonidae in North America, 53% are recorded from only one insect host species. Recent research suggests that natural enemies may not only be specific to a particular pest species, but it may be specific to only a particular situation (e.g. habitat or crop). When attacked by particular insect pests, some crops are now known to to produce very specific volatile chemicals which attract only particular natural enemies of that pest to that crop. Specificity is extremely important in biological control: on the one hand highly specific natural enemies are less likely to affect desirable non-target species (e.g. crops and beneficial insects) while, on the other hand, the high specificity of many natural enemies limits their commercial application by limiting the range of target species and hence markets.

Functionality is important in biological control for two reasons. Firstly, it helps us to understand how we might make use of the bewilderingly large diversity of natural enemies – it helps to sort natural enemies into functional groups which behave in a similar manner and which can be sampled, used or conserved in a similar manner. Secondly, it identifies a degree of redundancy in biological control systems. That is, it shows us how more than one natural enemy may make a contribution to control of a particular pest in a particular situation. This allows us to be efficient in selecting natural enemies to use in biological control. If we are re-creating a complex of natural enemies on a pest, for instance on an alien pest which has been introduced without its natural enemies, functionality allows us to be more strategic about putting together that complex.

At the same time, when natural enemy diversity is threatened in a crop system (e.g. by intensive cropping practices), functionality helps us understand how many and what kind of species need to remain in a system to ensure biological control remains effective. Where taxonomy is still very poor, as for instance in soil organisms where perhaps only a very small percentage of organisms can be isolated, cultured and named, we may base our understanding and use of natural enemies on different functional groups rather than taxonomic groups. Recent techniques for extracting and analysing genetic material from soils have created a breakthrough in our ability to characterise them biologically and functionally.

At this point, the following conclusions may be drawn about natural enemy biodiversity relevant to its conservation and use in agriculture. The use of natural enemies in biological control is made more difficult by their sheer diversity and our poor understanding of this. This contrasts greatly with plant genetic resources, and makes biological control very dependent on taxonomy and taxonomical services. Understanding specificity and functionality is crucial to dealing with this diversity and "extracting" useful species to consider for biological control, and this makes biological control very dependent on ecology and ecological research, to understand the complex, dynamic relations between natural enemies and pests.

Approaches to Biological Control

Natural enemies exert a continuous level of pest suppression and regulation in agro-ecosystems without any specific human intervention. This is a form of biological control. At the same time, the term "biological control" is also used to describe the deliberate application of specific natural enemies against specific pests. This usually involves their production and release, sometimes as commercial products.

Not surprisingly, much of the history of biological control has focused on its second interpretation, the active use of specific natural enemies, because this is most closely associated with solving particular pest "crises" or developing pest control products. Interest in biological control as a continuing, natural process of pest suppression has waxed and waned over the past two centuries, and has its origins in early ecological research in the 19th century which understood natural enemies as a balancing force acting on population growth and size. Interest in using specific natural enemies for controlling specific pests arose at the very end of the 19th century, and grew rapidly in the first half of the 20th century, as agriculture became more intensive and pest problems more severe. However, with the widespread introduction of synthetic organic pesticides after 1950, interest in biological control, both as an underpinning process in crop production and as a technology, waned. Subsequently, resurgence of pests, associated with pesticide effects on natural enemies, led to a renaissance of interest in this "background" level of natural enemy action, whose conservation subsequently underpinned the concept of integrated pest management (IPM). The birth of this concept then stimulated a renaissance in biological control technology, aimed at developing IPM-friendly products to replace chemical pesticides.

Across this history, three general approaches to biological control arose. I will use definitions which are in line with current textbook treatments:

- *Natural enemy conservation* – the protection and encouragement of local natural enemy populations by crop and habitat management measures that enhance their survival, efficiency and growth.
- *Augmentation of natural enemies* – the release of natural enemies into crops to suppress specific populations of pests over one or a few generations, often involving the mass production of natural enemies.
- *Introduction of natural enemies* – the introduction locally of new species of natural enemies with the intention that they establish and build populations that suppress particular pests, often alien pests to which they are specific. This approach has frequently been called "classical biological control", probably because of its origins in the 19th century.

While these three approaches are distinctively different, they do interact in ways which important to natural enemy conservation.

Conservation of Natural Enemies

In natural ecosystems, predators, parasites and pathogens of animals, and herbivores of plants, contribute to the maintenance of complex communities at relatively stable levels. Outbreaks of particular species may occur in nature, but these tend to be suppressed over time by the density- and spatially-dependent, regulatory processes which characterise natural biological control. In agro-ecosystems, where we designate many species which limit crop production as pests, their control by natural enemies may be substantial, but may still fall short of what we anticipate in order to get the maximum productivity of the crops which we plant.

There is considerable evidence that agricultural intensification affects, and often reduces, levels of natural biological control. This is probably associated with the effects of intensification on cropassociated biodiversity. Along a gradient of intensification running from shifting cultivation and species-rich home gardens through to intensive cereal production, we see a dramatic change in biodiversity within crops, both above and below ground. This may be qualitative, for instance soil tillage and loss of accumulating litter tends to increase bacteria and small invertebrates relative to fungi and larger invertebrates, without always causing a shift in overall diversity. Or it may be quantitative, for instance, coffee grown under shade with mixed cropping has a diversity of ant and beetle communities similar to that of nearby rainforests, but this is greatly reduced in unshaded coffee monoculture.

Precisely how these changes in biodiversity affect biological control will depend on the functionality of the community remaining after intensification. It is not necessarily the case that intensification reduces the contribution of biological control. Rice ecosystems are examples of intensive production where an effective insect natural enemy complex is maintained. However, a range of modern farming practices have been shown to be directly antagonistic to biological control. Intensive systems based on genetically uniform, high-yielding crops and high inputs of fertiliser and pesticides, may favour pest population growth and interfere with natural enemy survival and growth, leaving biological control operating well below its potential. The conservation of natural enemies in agro-ecosystems is targeted at maximising the impact of local natural enemies on pests. With a knowledge of local natural enemy communities, conditions which optimise biological control can be selected. These may include maintenance of uncultivated crop margins, where natural enemies find shelter or food, or survive on other prey or hosts between crops or seasons. Similar refuges may be created in crops for natural enemies. Multi-cropping and inter-cropping have been shown to increase insect natural enemy diversity and abundance relative to that in intensive monocultures. This, along with the effect of diversified crops on directly reducing pest populations contributes to the phenomenon of reduced pest impact on diversified systems.

Recent studies of insect chemical ecology have revealed the capacity of different plants to repel pests and attract natural enemies, sometimes called a "push-pull" strategy, and this has been applield in Africa to reduce stem borer damage on maize by intercropping and companion planting. Less intensive cropping systems may be associated with a greater diversity of soil organisms of importance to biological control, although there may not be a clear relationship between intensification and soil biodiversity in general. Considerable research has been directed at conservation of suppressive soils, which appear to have a microflora and fauna which naturally suppresses plant diseases and soil pests. Typically for biological control, this discovery has led to more efforts to identify particular control agents for commercial development, than to understand the ecology of these complex suppressive systems, so as to manage crops to favour them. Nonetheless, suppressive soils as a conservation approach have much promise, particularly with the new molecular techniques for characterising the genetic diversity and function of soils.

More specific crop manipulations can be made to favour particular natural enemy species or groups, including encouragement of specific pollen and nectar-rich wildflowers along crop margins or application of "food sprays" for predators on crops. This approach has recently been called "ecological engineering" and requires a deep understanding of natural enemy ecology and their use of animal and plant foods inside and outside the crop. Sometimes we discover important aspects of natural enemy conservation only when we disrupt biological control. The use of pesticides, particularly insecticides, has often been associated with resurgence of arthropod pests through the selective reduction of their natural enemy communities.

Such pesticide-induced resurgence in crops as varied as cotton, tree fruit, vegetables and rice has stimulated the development of integrated pest management (IPM) systems. Similar processes, involving pesticidal disruption of natural enemy communities and actions, can be found in the management of crop diseases, and IPM methods have been extended into plant disease and weed control. IPM can be defined as "the best mix of control tactics for a given pest problem in comparison with yield, profit and safety of alternative mixes". But behind this broad and sensible definition is the original impetus provided by pesticide-induced pest resurgence, which started in the 1950s. Sometimes resurgence occurred in the target pest, but often it occurred in minor pests after insecticides were used to control other species. Research at the time came up with many explanations for resurgence, but the elimination of natural enemies was probably the most important. A widespread phenomenon was resurgence of sucking insects such as scales and mealybugs. Here, the pests are somewhat protected from sprays by their concealed habits and waxy coats, but their tiny natural enemies are highly susceptible as they pick up pesticide while searching plant surfaces for their concealed hosts.

Resurgence often led to increased pesticide application, hence even more resurgence and sometimes pest resistance to chemicals, leading farmers onto a "pesticide treadmills" of spiralling input costs and falling yields. In these circumstances, the first step in many IPM programmes was simply the reduction of pesticide use, which allowed natural enemies to recover, increased yield and reduced costs, often over a very short period. Ultimately, in many crop systems, including those not showing resurgence, the IPM "best mix" of control methods reflects this combination of natural enemy conservation and limited pesticide use.

While IPM systems in developed and developing countries have different features, all tend to treat natural enemy action as a "baseline" of pest control, the conservation of which may make intervention with pesticides unnecessary. When this natural biological control is insufficient, then pesticides or other interventions are used, in such a way so as not to disrupt natural enemy action. Monitoring of some kind is therefore necessary in IPM to ensure that natural enemies are present and other interventions are unnecessary. This examination is usually conducted by trained extensionists or farmers.

Hence, one of the challenges in natural enemy conservation is that it requires knowledge of natural enemies and their biology. Many farmers may not recognise the difference between insect natural enemies of insect pests or weeds and insect pests themselves, and may treat them all with insecticides. Similarly, farmers may not recognise pests which are suffering from disease caused by natural enemies, and spray unnecessarily. In many parts of the world, generations of dependence on chemical pesticides has left many farming communities without an appreciation or understanding of natural enemies and biological control, and hence an inability to make use of them in IPM. IPM programmes therefore involve training of extensionists and/or farmers in recognition of natural enemies and understanding of biological control, empowering them to interpret changing conditions in crops and making decisions whether intervention is necessary. The cost of such training and monitoring may constrain the implementation of IPM. Farmer training in natural enemy conservation and IPM has proven effective, a model for which has been the "farmer field school" developed in the 1980s for rice IPM in Asia.

IPM systems tend to be developed by governments, in response to pesticide-induced crises in production, affecting the livelihoods of farming communities, or threatening consumer safety or the environment. In many developing countries, national IPM programmes have often developed around crops in crisis, such as cotton, rice and vegetables, and international agencies, particularly FAO, have provided technical assistance to national and regional programmes.

Most development assistance agencies have modified their traditional pesticide procurement guidelines on loans and grants to encourage an IPM approach. In Europe, the IOBC has a long tradition of developing and disseminating IPM standards for crops, which have been used by the agriculture and food industries as a basis for certification standards. More recently a number of European countries have adopted pesticide reduction targets, while in the USA, a different approach has been taken, involving a government policy to put 75% of crop production under IPM. To build their natural enemy conservation elements, often from a zero baseline, all of these government IPM programmes rely heavily on agricultural research capacity in national institutes, universities and museums.

Other crop systems where IPM has proven very important in the past 50 years includes tree crops, such as apples, citrus, cocoa and coffee, where insecticide use has disrupted natural control of scale insects, mealybugs, leafminers and other moth pests and led to pesticide resurgence. IPM has also been successful in certain field crops like cotton and brassicas, where pesticides have induced outbreaks of sucking insect and moth pests. Hence, the conservation of natural enemies for pest management today involves first and foremost the recognition of natural enemies and their role, and their conservation in IPM systems. Beyond that, conservation and natural enemy action can be enhanced by adjustment of crops and cropping practices to favour persistence and growth of their populations.

Augmentation of Natural Enemies

In many agricultural systems, useful natural enemies may not persist, due to the nature of the habitat, crop practices or seasonal effects. Even if well established in agro-ecosystems, they may not reach densities capable of delivering sufficient pest control at the right time – pathogens of insects and plants, for instance, often require specific conditions and high host densities to cause self-spreading epidemics in pest populations. Augmentation addresses this problem through application of natural enemies to crops.

Two variations of augmentation are recognised, which are effectively extremes along a continuum. *Inundation* involves release of a population of natural enemies which is large enough to suppress a particular pest generation, with no expectation that they will reproduce and exert further control. *Inoculation*, on the other hand, involves release of smaller numbers of natural enemies to establish and reproduce on a pest population, building sufficient numbers over a few generations to suppress the pest population before it becomes damaging. Inoculation is often associated with highly seasonal systems, where early release of natural enemies ensures that pest populations do not escape biological control because of the late arrival of "wild" natural enemies into the crop in question. Inoculation places less demands on production, but is more knowledge intensive and unpredictable, and perhaps for this reason it is less popular. Micro-organisms (e.g. pathogens of pests) are often used inundatively and macro-organisms (e.g. insect predators and parasitoids) inoculatively.

The practice of augmenting local natural enemy action by releasing natural enemies in crops is ancient. In the 19th and early 20th century the practice began to grow as a modern pest management option along two paths. Some insect pathogens had a saprophytic habit which allowed them to be produced by fermentation on inexpensive substrates, such as agar or nutrient broths. This allowed mass production of insect pathogenic fungi and bacteria. The first experimental mass-released pathogen, or "biopesticide" was the fungus, *Metarhizium anisopliae,* mass produced in the 1880s against beetle pests.

A second approach involved the use of arthropod (insect and mite) natural enemies for control of arthropod pests. These augmentation methods often grew out of unsuccessful efforts at introduction, when introduced natural enemies failed to establish. Mass production and release of predators and parasitoids became feasible when inexpensive ways to produce their prey or hosts were developed. The most important development involved egg parasitoids of the genus *Trichogramma,* for control of moth pests. Here, the eggs of meal moths were used for mass production. They could be cheaply produced on surplus or waste grain. This practice of using cheaply produced alternate or "factitious" hosts made augmentation more affordable and competitive. Today, augmentative approaches to biological control remain largely restricted to use of pathogens, predators and parasitoids for controlling invertebrate pests. Glasshouse systems in Europe use a wide variety of macrobial products in the form of predators and parasitoids, each product associated with the control of one or just a few insect or mite pest species. Augmentation of predators and parasitoids is also practiced in the field, particularly in orchard systems and, in the tropics, in sugar cane. *Trichogramma* is still applied against moth pests in a wide range of field crops.

For microbial products, bacteria, viruses and fungi against insect pests dominate the marketplace. Microbial products, largely fungi, have also been developed for control of plant pathogens, particularly soil competitors and antagonists, as well as "mycoherbicides" for weed control.

Most biopesticides are specialised products for particular pests, but two exceptions are Bt, which is used against a wide range of Lepidoptera, Diptera and Coleoptera, and *Trichoderma* spp., which are used to suppress a range of soil inhabiting plant pathogens. Worldwide today, about 150 species of natural enemies are sold as macrobial and microbial products for insect pest control, but just 30 of these species account for 90% of sales. Augmentation today is a mixture of public and private sector activity. Public sector activity grew around state programmes, particularly in state-owned production systems such as existed in China and the former Soviet Union. Many of these have now been privatised locally.

Privatised production has long been a tradition in a number of crop systems, like citrus, glasshouses and sugar cane, where centralised production is efficient. In the case of sugar production, local milling companies may produce natural enemies and sell them to growers, while in the glasshouse industry in Europe, a few growers began to product natural enemies for their own use and gradually grew into major companies producing natural enemies for growers throughout the region. Because of this mix of public and private sector activity it is difficult to estimate the scale of augmentation today. From a commercial perspective, sales are in the order of $300-400m, which is only about 1% of the approximately $30-40b market in pest control products, the rest being sales of chemical pesticides and, increasingly, transgenic crops.

Expectations in the last two decades that commercial biological control products would displace a substantial proportion of chemical pesticide sales, which led to much venture capital investment, are now gone. Instead, augmentation is seen today to service a stable, important, but small set of niche markets in pest control. Commonly cited reasons for biological control products not having a larger market share are the high specificity of some biological products (which limits markets), short shelflife, problems in storage and transport, unpredictability in the field due to variable conditions on release (e.g. temperature, solar radiation, rainfall) and cost of production relative to alternative, chemical pesticides.

Underlying these shortcomings, however, is a pest control environment that has been developed around a "chemical paradigm" over the past 50 years, into which biological products fit poorly. This paradigm has favoured products with high efficacy (killing effect) and rapid action, which the market now expects. Chemicals generally out-perform biological products on these criteria, while properties that give biological products their advantage, namely their capacity to reproduce, spread and protect other natural control, have not been valued in conventional pest control markets. It is not surprising, perhaps that *Bacillus thuringiensis*, which accounts for over 90% of microbial biopesticide sales, is usually formulated as dead bacteria containing the toxic protein endotoxin – as such it is not a living biological control agent, but merely a chemical pesticide. Augmentation has found its niche on crops where chemical pesticides are undesirable, difficult to use and/or relatively ineffective. Examples include high value horticulture production, e.g. of vegetables, fruits and flowers, and some tropical plantation systems, like sugarcane. An interesting pattern of development has emerged in some augmentative programmes, best illustrated by European glasshouse production. Here, up to the 1950s growers relied on synthetic pesticides for pest control, until resistance developed in the alien spider mite, *Tetranychus urticae*.

Faced with no chemical control options, biological control was developed using an alien predatory mite, *Phytoseiulus persimilis*, discovered in Europe in 1960, and by the 1970s many growers were using this. Also in the 1970s, pesticide problems with control of an alien whitefly, *Trialeurodes vaporariorum* were addressed similarly with an alien parasitoid, *Encarsia formosa*. With these two biological control agents firmly established, subsequent new pests could not be treated with broadspectrum pesticides without disrupting existing control. Without pesticides, some secondary pests became more important. IPM programmes were established which used the least disruptive pesticides, and biological control agents were found and developed for these new pests, including thrips, aphids, leaf miners. As of 2004, there were about 115 natural enemies used commercially in these systems. To some extent, this glasshouse production system has become locked into a cascading of new biological control products and IPM methods, as use of broad-spectrum pesticides becomes less feasible. Concern about pesticide residues in glasshouse crops has also contributed to keeping biological control popular, and its use has facilitated other developments, such as the introduction of bumblebees to increase pollination of tomatoes.

Opportunities for such "cascading" biological control probably exist in other crop systems presently dominated by pesticides. Because augmentation is founded on a need to repeatedly introduce natural enemies into crop systems, its success has often required that it be commercialised. Today, it is increasingly a private sector activity, practiced either by companies which supply products to farmers or by farmers or farming cooperatives which produce natural enemies in a central facility and distribute or sell them to members. Multinational agrochemical corporations made an investment in biological control products in the last few decades of the 20th century. Few lasting products emerged from this, probably because they competed poorly with chemical products from the same companies. Experience suggests that the most sustainable biological control production for augmented pathogens and invertebrate natural enemies comes from small to medium scale enterprises, often serving a regional market and particular crop systems. Many of these private producers have developed from farming operations which initially produced products for local use. Universities and government research institutions have provided critical research support to these commercial operations.

In principle, developing countries should provide excellent conditions for augmentation, because of relatively high local pesticide costs and risks, a rich biological control resource and inexpensive labour and materials. However, sustainable augmentation requires also stable local demand and supply, a favourable regulatory system and a very good research backup, and these have often been lacking in developing countries.

Introduction of Natural Enemies

The international movement of peoples and crops has contributed for centuries to the introduction of new pests, diseases and weeds to agriculture. Often, introduced pests will exhibit devastating outbreaks, which may spread rapidly across a country. Given that many individual farm businesses are affected, governments usually step in with campaigns of pest eradication or control. Biological control introduction may be employed in these circumstances, particularly if it has been used effectively elsewhere against the pest.

Biological control introductions against alien agricultural insect pests emerged in the 19th century, stimulated by growing agricultural research in regions settled by Europeans where introduced pests on introduced crops were major problems. The first major success in biological control introduction was made in 1888, against the alien cottony cushion scale, *Icerya purchasi*, in California. The introduction of an Australian parasitoid and ladybird from the pest's area of origin was so successful that it stimulated similar introductions of ladybirds and parasitoids against other alien pests, and a long history of programmes began. In the early 1900s, the method was extended to the introduction of herbivores for the control of exotic weeds, like cactus in Australia in 1925. Greathead and Greathead and Julien and Griffiths have established databases on introduction programmes against insects and weeds, respectively, and have shown a steady increase in numbers of agents released per decade over the 20th century.

Most introduction programmes have been directed against alien insect pests or weeds. For both targets, insect natural enemies have been the preferred agents, largely because of their high host specificity and capacity for rapid reproduction and spread. Introduction programmes have been successfully implemented over a wide range of crops and habitats, but they have been most frequent against insect pests of perennial crops, such as tree crops, and against weeds of extensive habitats, such as pastures and waterways. Several factors have contributed to this bias. For insect pests, the establishment and spread of natural enemies may be more effective in less disrupted agro-ecosystems, such as orchards or forest plantations, than in field crops. Common alien insect pests, such as sucking insects and leaf miners, are more common in these systems.

Similarly, weeds of pastures, plantations and waterways may provide a more stable environment for establishment and spread of biological control agents than those of arable crops. However, in both cases, competition with chemical pest control has probably also been a factor. Biological control has found a niche where chemical control is prohibitively expensive, such as in extensive pastures, ineffective, such as in dense forests or sugarcane plantations, or undesirable, such as in waterways where pollution is a concern. However, it is important to stress that biological control introductions have wide applicability, and there have been spectacular successes for insect and weed control in annual, arable crops. In 1971, a specific strain of rust fungus gave widespread control of a weed of field crops, *Chondrilla juncea*, in Australian field crops, and heralded the start of the use of rust fungi for biological control.

Success rates of biological control for natural enemies which have been established in areas of introduction have been estimated at about 40% for programmes against insect pests and 30% for programmes against alien weeds. Success rate of all introduced agents would be lower, as some will fail to establish.

Introduction programmes require considerable research and time, hence they have high initial costs, but their benefits, where successful, are continuing. A typical programme involves a number of stages of research. First, studies are made to determine the area of origin of the alien pest, which often requires detailed taxonomy, as the target species may not be a pest in that region and may even be unknown to science. Exploration is made in the likely area of origin and potential natural enemies for introduction identified. These are characterised and studied for their potential impact on the pest in its new range and for their specificity. Safety protocols usually require that only highly specific natural enemies are introduced, eliminating a large number of species from consideration. Taxonomic support is critical throughout this process. On more than one occasion, misidentification of target pests and/or their natural enemies has caused delay or failure in introduction programmes.

Specificity tests follow well established protocols and may be lengthy, depending on the number of non-target species to be tested. This is particularly true for weed control, where a wide range of crop and native plants are usually tested. Selected natural enemies are imported into quarantine facilities in the affected country (or into similar facilities in cooperating countries) where tests are made to ensure they are carrying no contaminants themselves. They are then mass reared and released into affected areas, and their establishment and impact on the target pest monitored. A programme against a single alien pest may involve the introduction of several natural enemies, which may take years, even decades.

A concern for environmental impact of biological control introductions has become particularly important in the past 20 years and has led to more extensive safety testing of potential agents, particularly against native species.

Introduction programmes have been almost universally government activities, for several reasons. A new alien pest, weed or disease has the capacity to spread over many farms and may indeed be well established and spread before it is discovered. Hence, collective action is necessary to control it and stop further spread, which cannot easily be done by individual farmers. Further, introduction programmes are usually prohibitively expensive for farmers to mount. They require the deliberate introduction of alien species, which must be carefully regulated by government. Finally, the benefits of successful introduction spread as the introduced natural enemies spread, and cannot be limited to the farms of any private investors.

EFFECTS OF BIOLOGICAL CONTROL ON BIODIVERSITY

Effects on Native Biodiversity

Biological control can potentially have positive and negative effects on biodiversity. Most of the time a biological control is introduced to an area to protect a native species from an invasive or exotic species that has moved into its area. The control is introduced to lessen the competition among native and invasive species. However, the introduced control does not always target only the intended species. It can also target native species. When introducing a biological control to a new area, a primary concern is the host- or prey-specificity of the control agent.

Generalist feeders often make poor biological control agents, and may become invasive species themselves. For this reason, potential biological control agents should be subject to extensive testing and quarantine before release into any new environment. If a species is introduced and attacks a native species, the biodiversity in that area can change dramatically.

When one native species is removed from an area, it may have filled an essential niche, When this niche is absent it may directly affect the entire ecosystem. Because they tend to be generalist feeders, vertebrate animals seldom make good biological control agents, and many of the classic cases of "biocontrol gone awry" involve vertebrates. For example, the cane toad was introduced as a biological control and had significant negative impact on biodiversity.

The cane toad was intentionally introduced to Australia to control the cane beetle. When introduced, the cane toad thrived very well and did not only feed on cane beetles but other insects as well. The cane toad soon spread very rapidly, thus taking over native habitat. The introduction of the cane toad also brought foreign disease to native reptiles. This drastically reduced the population of native toads and frogs. "The cane toad also exudes and can squirt poison from the parotid glands on their shoulders when threatened or handled. This toxin contains a cocktail of chemicals that can kill animals that eat it.

Freshwater crocodiles, goannas, tiger snakes, dingos and northern quolls have all died after eating cane toads, as have pet dogs ". This goes to show a small but deadly organism can alter the native biodiversity in an ecosystem in a very expedient manner. A pyramid effect can take place if native species are reduced or eradicated. The domino effect keeps on going and can potentially exude on other bordering ecosystems until an equilibrium is reached.

A second example of a biological control agent that subsequently crossed over to native species is the Rhinocyllus conicus. The seed feeding weevil was introduced to

North America to control exotic thistles. However, the weevil did not target only the exotic thistles, it also targeted native thistles that are essential to various native insects.

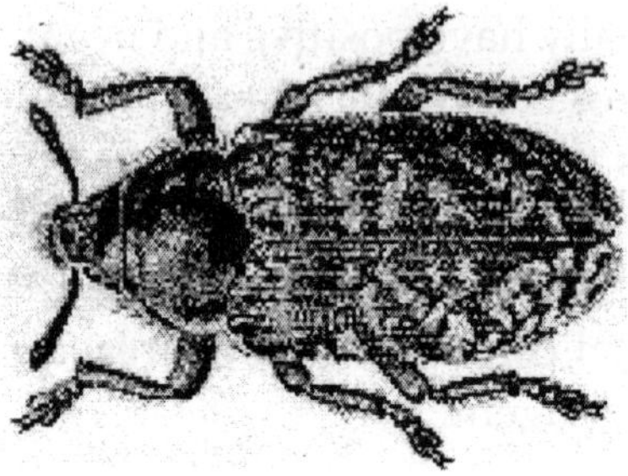

Figure 1. Rhinocyllus conicus

The native insects rely solely on native thistles and do not adapt to other plant species. Therefore, they cannot survive. Biological controls do not always have negative impacts on biodiversity. Successful biological control reduces the density of the target species over several years, thus providing the potential for native species to re-establish. In addition, regeneration and reestablishment programs can aid to the recovery of native species.

Native species can be affected in a positive way as well. To develop or find a biological control that exerts control only on the targeted species is a very lengthy process of research and experiments. In the late 1800's, the citrus industry was in great fear when the cottony cushion scale was discovered. This organism could cause a great deal of economic loss to the industry. However, a biological control was introduced.

The vedalia beetle and a parasitoid fly were introduced to control the pest. Within a few years time, the cottony cushion scale was controlled by the natural enemies and the citrus industry suffered little financial loss. Many exotic or invasive species can suppress the development of native species. The introduction of an effective biological control that reduces the population of the invasive species allows the rejuvenation of the native species. Biological controls can reduce competition for biotic and abiotic factors which can result in the re-establishment of the once over ran native species.

Effects on Invasive Species

Invasive species are closely associated with biological controls because the environment in which they are invasive most likely does not contain their natural enemies. If invasive species are not controlled, biodiversity may be at great threat in the affected area. An example of an invasive species is the alligator weed. This plant was introduced to the United States from South America. This aquatic weed spreads very rapidly and causes many problems in lakes and rivers. The weed takes root in shallow water causing major problems such as navigation, irrigation, and flood control. The alligator weed flea beetle and two other biological controls were released in Florida. Because of their success, Florida banned the use of herbicides to control alligator weed three years after the controls were introduced. Biological controls for invasive species also can have a negative impact on biodiversity. The cane toad, as mentioned previously, is a great example of trying to control an invasive species.

The cane toad was introduced to eradicate an invasive species. However, it became invasive, thus altering the biodiversity. The introduction of the cane toad could have potentially caused more of a disturbance in biodiversity than the targeted species did.

Effects on Future

With further research and more scientific experiments, biological control could potentially play a huge role in the future of pest prevention. Biological control is being used among society today; however, it could someday reduce the use of many pesticides and herbicides. Since biological control could potentially have a large economic value, if found to be successful, research and job fields would increase continually. By increasing awareness of biological controls among more people, new successful biological controls could be discovered in the future. This could eliminate the overuse of chemicals. Biodiversity would increase because untargeted species that are exterminated with chemicals would no longer occur.

Economic Effects

Therefore, biological control is heavily analysed by the amount of economic gain that directly comes from biological control. Many of the known economics of biological control are related directly to agriculture practices. Since agriculture has a huge impact on biodiversity this could potentially increase the biodiversity among agricultural practices.

In order for agriculture to keep up with the growing population, many inputs are increased resulting in the loss of un-harmful species. Biological control use has been very minimal in agriculture. Less than 1% of global pest control sales of $30 billion involve biological methods. Very few case studies on the cost-benefit analysis of biological control have been done however a few have taken place.

A Critical evaluation of augmentative biological control has found four case studies. In one case, "the releases of a parasitoid Gryon pennsylvanicum Ashmead to control the true bug Anasa tristis DeGeer on pumpkins produced lower net benefit than applications of esfenvalerate (pesticide); 18% lower in one year and 120% lower in the next. In 1 year of the study, a combination of augmentative releases and use of a resistant pumpkin variety produced greater net benefit than pesticide alone, but not pesticide combined with the resistant variety".

Another case study found that "calculated that releases of *T. nubilale* were considerably less cost-effective than pesticide applications used to control ECB on feed corn and fresh-market sweet corn. Pesticide applications produced 87% and 45% more net benefit than augmentation for feed corn and fresh market corn, respectively. In seed corn, however, *Trichogramma* releases produced essentially equivalent net benefits to pesticide treatments. In a third cost-benefit analysis of augmentation, Lundgren et al. showed that Trichogramma brassicae Bezdenko releases produced considerably less net benefit than methomyl treatments.

CLASSICAL BIOLOGICAL CONTROL

Classical biological control is the introduction of natural enemies to a new locale where they did not originate or do not occur naturally. This is usually done by government authorities. In many instances the complex of natural enemies associated with an insect pest may be inadequate. This is especially evident when an insect pest is accidentally introduced into a new geographic area without its associated natural enemies. These introduced pests are referred to as exotic pests and comprise about 40% of the insect pests in the United States. Examples of introduced vegetable pests include the European corn borer (*Ostrinia nubilalis*), one of the most destructive insects in North America. To obtain the needed natural enemies, scientists turned to classical biological control. This is the practice of importing, and releasing for establishment, natural enemies to control an introduced (exotic) pest, although it is also practiced against native insect pests.

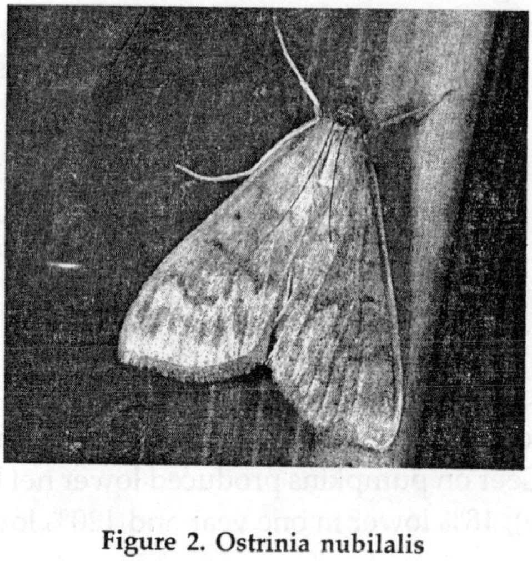

Figure 2. Ostrinia nubilalis

The first step in the process is to determine the origin of the introduced pest and then collect appropriate natural enemies associated with the pest or closely related species. The natural enemy is then passed through a rigorous quarantine process, to ensure that no unwanted organisms (such as hyperparasitoids) are introduced, then they are mass produced, and released. Follow-up studies are conducted to determine if the natural enemy becomes successfully established at the site of release, and to assess the long-term benefit of its presence.

There are many examples of successful classical biological control programs. One of the earliest successes was in controlling Icerya purchasi, the cottony cushion scale, a pest that was devastating the California citrus industry in the late 1800s. A predatory insect Rodolia cardinalis, and a parasitoid fly were introduced from Australia. Within a few years the cottony cushion scale was completely controlled by these introduced natural enemies.

Damage from Hypera postica Gyllenhal, the alfalfa weevil, a serious introduced pest of forage, was substantially reduced by the introduction of several natural enemies. About 20 years after their introduction, the population of weevils, in the alfalfa area treated for alfalfa weevil in the Northeastern United States, was reduced by 75 percent.

A small wasp, Trichogramma ostriniae, was introduced from China to help control the European corn borer making it a recent example of a long history of classical biological control efforts for this major pest. Many classical biological control programs for insect pests and weeds are under way across the United States and Canada. The population of Levuana irridescens, a serious coconut pest in Fiji was brought under control by a classical biological control program in the 1920s.

Classical biological control is long lasting and inexpensive. Other than the initial costs of collection, importation, and rearing, little expense is incurred. When a natural enemy is successfully established it rarely requires additional input and it continues to kill the pest with no direct help from humans and at no cost. Unfortunately, classical biological control does not always work. It is usually most effective against exotic pests and less so against native insect pests. The reasons for failure are not often known, but may include the release of too few individuals, poor adaptation of the natural enemy to environmental conditions at the release location, and lack of synchrony between the life cycle of the natural enemy and host pest.

AUGMENTATION

This type of biological control involves the supplemental release of natural enemies. Relatively few natural enemies may be released at a critical time of the season or literally millions may be released. Additionally, the cropping system may be modified to favor or augment the natural enemies. This latter practice is frequently referred to as habitat manipulation.

An example of inoculative release occurs in greenhouse production of several crops. Periodic releases of the parasitoid, Encarsia formosa, are used to control greenhouse whitefly, and the predaceous mite, Phytoseiulus persimilis, is used for control of the two-spotted spider mite.

Lady beetles, lacewings, or parasitoids such as those from the genus Trichogramma are frequently released in large numbers. Recommended release rates for Trichogramma in vegetable or field crops range from 5,000 to 200,000 per acre per week depending on level of pest infestation. Similarly, entomopathogenic nematodes are released at rates of millions and even billions per acre for control of certain soil-dwelling insect pests.

Habitat or environmental manipulation is another form of augmentation. This tactic involves altering the cropping system to augment or enhance the effectiveness of a natural enemy. Many adult parasitoids and predators benefit from sources of nectar and the protection provided by refuges such as hedgerows, cover crops, and weedy borders. Also, the provisioning of natural shelters in the form of wooden caskets, boxes or flowerpots is a form of this. For example, the stimulation of the natural predator Dermaptera is done in gardens by hanging up turnaround flowerpots with straw or wood wool.

Mixed plantings and the provision of flowering borders can increase the diversity of habitats and provide shelter and alternative food sources. They are easily incorporated into home gardens and even small-scale commercial plantings, but are more difficult to accommodate in large-scale crop production.

There may also be some conflict with pest control for the large producer because of the difficulty of targeting the pest species and the use of refuges by the pest insects as well as natural enemies. Examples of habitat manipulation include growing flowering plants such as Buckwheat near crops to attract and maintain populations of natural enemies. For example, hover fly adults can be attracted to umbelliferous plants in bloom. Biological control experts in California have demonstrated that planting prune trees in grape vineyards provides an improved overwintering habitat or refuge for a key grape pest parasitoid.

The prune trees harbor an alternate host for the parasitoid, which could previously overwinter only at great distances from most vineyards. Caution should be used with this tactic because some plants attractive to natural enemies may also be hosts for certain plant diseases, especially plant viruses that could be vectored by insect pests to the crop. Although the tactic appears to hold much promise, only a few examples have been adequately researched and developed.

Examples of Predators

Ladybugs, and in particular their larvae which are active between May and July in the northern hemisphere, are voracious predators of aphids such as greenfly and blackfly, and will also consume mites, scale insects and small caterpillars. The ladybug is a very familiar beetle with various colored markings, whilst its larvae are initially small and spidery, growing up to 17 mm long.

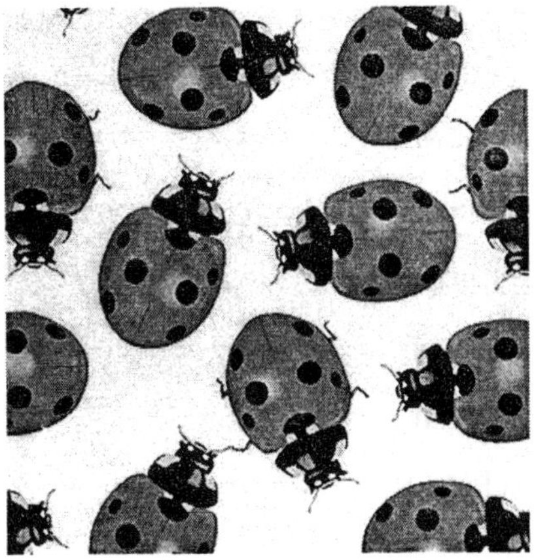

Figure 3. Ladybugs

The larvae have a tapering segmented grey/black body with orange/yellow markings and ferocious mouthparts. They can be encouraged by cultivating a patch of nettles in the garden and by leaving hollow stems and some plant debris over winter so that they can hibernate. Hoverflies resemble slightly darker bees or wasps and they have characteristic hovering, darting flight patterns. There are over 100 species of hoverfly whose larvae principally feed upon greenfly, one larva devouring up to fifty a day, or 1000 in its lifetime. They also eat fruit tree spider mites and small caterpillars.

Adults feed on nectar and pollen, which they require for egg production. Eggs are minute (1 mm), pale yellow white and laid singly near greenfly colonies. Larvae are 8-17 mm long, disguised to resemble bird droppings, they are legless and have no distinct head. Semi-transparent in a range of colours from green, white, brown and black.

Hoverflies can be encouraged by growing attractant flowers such as the poached egg plant *(Limnanthes douglasii)*, marigolds or phacelia throughout the growing season. Dragonflies are important predators of mosquitoes, both in the water, where the dragonfly naiads eat mosquito larvae, and in the air, where adult dragonflies capture and eat adult mosquitoes. Community-wide mosquito control programs that spray adult mosquitoes also kill dragonflies, thus removing an important biocontrol agent, and can actually increase mosquito populations in the long term. Other useful garden predators include lacewings, pirate bugs, rove and ground beetles, aphid midge, centipedes, predatory mites, as well as larger fauna such as frogs, toads, lizards, hedgehogs, slow-worms and birds. Cats and rat terriers kill field mice, rats, June bugs, and birds. Dogs chase away many types of pest animals. Dachshunds are bred specifically to fit inside tunnels underground to kill badgers.

- *Phytoseiulus persimilis* (against spider mites)
- *Amblyseius californicus* (against spider mites)
- *Amblyseius cucumeris* (against spider mites)
- *Typhlodromips swirskii* (against spider mites, thrips, and white flies)
- *Feltiella acarisuga* (against spider mites)
- *Stethorus punctillum* (against spider mites)
- *Macrolophus caluginosus* (against spider mites)
- *Encarsia formosa* (against white flies)
- *Eretmocerus* spp. (against white flies)

Parasitoid Insects

Most insect parasitoids are wasps or flies. Parasitiods comprise a diverse range of insects that lay their egg on or in the body of an insect host, which is then used as a food for developing larvae. Parasitic wasps take much longer than predators to consume their victims, for if the larvae were to eat too fast they would run out of food before they became adults. Such parasites are very useful in the organic garden, for they are very efficient hunters, always at work searching for pest invaders. As adults they require high energy fuel as they fly from place to place, and feed upon nectar, pollen and sap, therefore planting plenty of flowering plants, particularly buckwheat, umbellifers, and composites will encourage their presence.

Four of the most important groups are:

- *Ichneumonid wasps:* (5-10 mm). Prey mainly on caterpillars of butterflies and moths.
- *Braconid wasps:* Tiny wasps (up to 5 mm) attack caterpillars and a wide range of other insects including greenfly. A common parasite of the cabbage white caterpillar- seen as clusters of sulphur yellow cocoons bursting from collapsed caterpillar skin.
- *Chalcid wasps:* Among the smallest of insects (<3 mm). Parasitise eggs/larvae of greenfly, whitefly, cabbage caterpillars, scale insects and Strawberry Tortrix Moth.

- *Tachinid flies:* Parasitise a wide range of insects including caterpillars, adult and larval beetles, true bugs, and others.

Figure 4. Ichneumonid wasps

BIOLOGICAL CONTROL WITH MICRO-ORGANISMS

Various microbial insect diseases occur naturally, but may also be used as biological pesticides. When naturally occurring, these outbreaks are density dependent in that they generally only occur as insect populations become denser.

Bacteria and Biological Control

Bacteria used for biological control infect insects via their digestive tracts, so insects with sucking mouth parts like aphids and scale insects are difficult to control with bacterial biological control. Bacillus thuringiensis is the most widely applied species of bacteria used for biological control, with at least four sub-species used to control Lepidopteran (moth, butterfly), Coleopteran (beetle) and Dipteran (true flies) insect pests.

Fungi and Biological Control

Fungi that cause disease in insects are known as entomopathogenic fungi, including at least fourteen species of entomophthoraceous fungi attack aphids. Species in the genus Trichoderma are used to manage some soilborne plant pathogens.

PLANTS TO REGULATE INSECT PESTS

Choosing a diverse range of plants for the garden can help to regulate pests in a variety of ways, including;

- Masking the crop plants from pests, depending on the proximity of the companion or intercrop.
- Producing olfactory inhibitors, odors that confuse and deter pests.
- Acting as trap plants by providing an alluring food that entices pests away from crops.
- Serving as nursery plants, providing breeding grounds for beneficial insects.
- Providing an alternative habitat, usually in a form of a shelterbelt, hedgerow, or beetle bank where beneficial insects can live and reproduce. Nectar-rich plants that bloom for long periods are especially good, as many beneficials are nectivorous during the adult stage, but parasitic or predatory as larvae. A good example of this is the soldier beetle which is frequently found on flowers as an adult, but whose larvae eat aphids, caterpillars, grasshopper eggs, and other beetles.

The legume vine Mucuna pruriens is used in the countries of Benin and Vietnam as a biological control for problematic Imperata cylindrica grass. *Mucuna pruriens* is said not to be invasive outside its cultivated area.

DIRECT INTRODUCTION OF BIOCONTROL AGENTS

Most of the biological controls listed above depend on providing incentives in order to 'naturally' attract beneficial insects to the garden. However there are occasions when biological controls can be directly introduced. Common biocontrol agents include parasitoids, predators, pathogens or weed feeders. This is particularly appropriate in situations such as the greenhouse, a largely artificial environment, and are usually purchased by mail order. Some biocontrol agents that can be introduced include;

- *Encarsia formosa.* This is a small predatory chalcid wasp which is parasitical on whitefly, a sap-feeding insect which can cause wilting and black sooty moulds. It is most effective when dealing with low level infestations, giving protection over a long period of time. The wasp lays its eggs in young whitefly 'scales', turning them black as the parasite larvae pupates. It should be introduced as soon as possible after the first adult whitefly are seen. Should be used in conjunction with insecticidal soap.
- Red spider mite, another pest found in the greenhouse, can be controlled with the predatory mite *Phytoseilus persimilis*. This is slightly larger than its prey and has an orange body. It develops from egg to adult twice as fast as the red spider mite and once established quickly overcomes infestation.
- A fairly recent development in the control of slugs is the introduction of 'Nemaslug', a microscopic nematode (*Phasmarhabditis hermaphrodita*) which will seek out and parasitise slugs, reproducing inside them and killing them. The nematode is applied

by watering onto moist soil, and gives protection for up to six weeks in optimum conditions, though is mainly effective with small and young slugs under the soil surface.

- A bacterial biological control which can be introduced in order to control butterfly caterpillars is *Bacillus thuringiensis*. This available in sachets of dried spores which are mixed with water and sprayed onto vulnerable plants such as brassicas and fruit trees. The bacterial disease will kill the caterpillars, but leave other insects unharmed. There are strains of Bt that are effective against other insect larvae. *Bt israelensis* is effective against mosquito larvae and some midges.

The European Rabbit (*Oryctolagus cuniculus*) is seen as a major pest in Australia

- A viral biological control which can be introduced in order to control the overpopulation of European rabbit in Australia is the rabbit haemorrhagic disease virus that causes the rabbit haemorrhagic disease.
- A biological control being developed for use in the treatment of plant disease is the fungus Trichoderma viride. This has been used against Dutch Elm disease, and to treat the spread of fungal and bacterial growth on tree wounds. It may also have potential as a means of combating silver leaf disease.
- The parasitoid Gonatocerus ashmeadi (Hymenoptera: Mymaridae) has been introduced to control the glassy-winged sharpshooter *Homalodisca vitripennis* (Hemipterae: Cicadellidae) in French Polynesia and has successfully controlled ~95% of the pest density.

BIOLOGICAL CONTROL IN IPM SYSTEMS

The classification of biological control into different approaches is arbitrary, and these different approaches are strongly inter-linked. For instance, augmentation and introduction approaches lie along a continuum of persistence of natural enemies in an agro-ecosystem. Introduction of a particular natural enemy may work in certain circumstances, while augmentation of the same species may be necessary in another, because its persistence may be reduced or only seasonal. Further, once natural enemies are released in the environment, for augmentation or introduction, they become part of the local natural enemy community. Hence, natural enemy conservation is an important part of successful augmentation and introduction: there is little value to releasing natural enemies into a system that is using pesticides in a way that will kill them off. Augmentation measures are often deliberate elements of IPM systems. When local natural enemies do not provide sufficient pest control, augmentation may be an effective intervention that avoids the disruptive effect of chemical pesticides. Similarly, alien pests disrupt local pest control systems. Historically, pesticide use against new alien pests often causes resurgence of secondary, local pest species. Introduction of specific and effective natural enemies may suppress these pests, eliminate pesticide use and restore IPM systems. IPM in many tree crop systems (e.g. fruit, nuts, forestry) is underpinned by former introduction programmes against alien pests such as scale insects, aphids, mealybugs. In these systems, as with the cascade systems of protected cultivation, maintaining successful biological control is so important to production that new pest problems require non-chemical, usually biological solutions – hence introduction programmes may *drive* biological control into crop systems.

In IPM systems, biological control interacts with other control measures. Chemical pesticides have a history of antagonism with biological control, but in recent decades, implementation of thresholdbased pesticide use of IPM and the development of more selective chemical products has greatly improved opportunities for integrating biological and chemical control. Further, biological control has a critical role to play in maintaining the availability of necessary pesticides in IPM systems. Pest resistance to pesticides is widespread, and evolves and spreads particularly rapidly when intensive pesticide use causes strong selection pressure on pest populations. Natural enemies, by reducing both pesticide use and pest populations, slow this evolutionary process. The introduction of IPM, based on natural enemy conservation not only reduced pest populations but restored the contribution of resistant plant varieties to pest suppression.

Another key element of IPM, the breeding or engineering of plants resistant to pests, has strong potential compatibility with biological control. Ecological theory suggests that successful biological control requires natural enemies to have the efficiency and reproductive capacity to overcome pest population growth. Plant resistance reduces the growth rate of pest populations, and hence would facilitate the successful action of natural enemies. It is likely that the complementary nature of plant resistance and biological control operates in many crop systems. In practice, however, the deliberate efforts of plant breeders and biological control practitioners have usually each sought their own "single technology solutions" to pest problems and integration has been limited. Because there is so much "partial resistance" created in plant breeding which remains under-utilised, and so many biological control opportunities, through conservation or augmentation, that provide good but not complete or continuous control, the capacity to integrate partial plant resistance and biological control to provide complete control of pests is considerable.

Value of Biological Control

The justification for conservation and use of biodiversity for biological control must rest on demonstrating the value of biological control to agriculture and society. To value biological control, we might measure its contribution to the level of production of food, fibre and other crop products and/or to the stability or security of that production. However, biological control may have indirect benefits as well, for instance its value in reducing environmental effects of excessive pesticide use and contributing generally to sustainable food production systems. Further, biological control approaches may have application beyond agriculture, for instance, in the control of alien weeds in national parks, or in the control of insect vectors of infectious human or animal diseases.

Efforts to place an economic value on biological control have been patchy. Studies are often either very narrow, focusing on one pest in one control programme, and therefore difficult to extrapolate, or very broad and based on many assumptions, hence speculative. Entirely different methods have been used to value conservation, augmentation and introduction approaches. For this reason, no attempt is made here to provide a consolidated valuation, rather valuation of each approach will be reviewed and these will then be integrated and compared.

A number of attempts have been made to value biological control as an ecosystem service. These calculations have been at such a general level that they may usually be considered to incorporate the value of all approaches to biological control, including conservation, augmentation and introduction of natural enemies. In most studies, a range of services from biodiversity are estimated, with biological control being just one of these. Studies of this kind often make very broad and challenging assumptions.

Costanza *et al.* estimate a total value of ecosystems services worldwide of $33 trillion annually and, without presenting detail on calculations, attribute $417b of this to biological control. However, most of the value of this service was attributed to natural ecosystems, mostly marine, only $34b was attributed to cropland itself. Naylor and Ehrlich derive a strictly agricultural value for biological control in the USA by combining the cost of pesticides and their subsidies, plant breeding for resistance and pest control externalities and postulating that biological control, if widely used, would offset much of this. From this they estimate the value of biological control to the US as between $54b and $1 trillion per annum. However, while pesticide use does sometimes only replace natural control with chemical control, it is not realistic to assume that biological control would completely replace chemical control and plant resistance, at no cost.

Pimentel *et al.* and Pimentel assumed that natural enemies provide about 60% of the non-pesticidal control of pests. In the USA, they estimate that this is worth $12b per annum (in contrast to $20b benefits from pesticidal control) and worldwide $100b. Losey *et al.* have used similar methods for estimating the value of *insect* natural enemies in the USA. They assumed that natural control (including natural enemies, crop resistance and environmental effects) suppresses 65% of the potential damage to crops in the USA from insect pests, and calculate that this saves $14b to agriculture annually, $4.5b of which is attributable to insect natural enemies.

Our understanding of the value of natural enemy conservation is based almost entirely on the comparison of agro-ecosystems with and without natural enemies. Three lines of evidence indicate that, without the self-renewing action of local natural enemies, many pest species, or potential pest species, would be more abundant, causing substantial losses:

- Pest resurgence, where pesticides are shown to eliminate local natural enemies while having limited effect on the target pest, leading to pest outbreaks and losses.
- Biological control introductions, which reverses this observation, demonstrating that addition of specific natural enemies to pests which have escaped natural control typically reduces their populations, often by two or more orders of magnitude.
- Experimental field studies which physically or chemically exclude natural enemies from pest populations and record substantial pest population increases.

Putting an economic value on this reduction of pest burden has been achieved in some IPM programmes where reducing pesticide use has restored control by natural enemies. Van den Berg has reviewed 25 IPM programmes based on farmer field schools, all of which involved reduction of pesticide use and conservation of natural enemies. Across these programmes, based in Asia, Africa and Latin America, largely on rice, cotton, vegetables, there was a consistent and usually dramatic reduction in pesticide use in the year following training, and often an increase in yield.

Economic benefits to farmers were widespread, but only one study related benefits to costs of training. This study, on rice IPM in Sri Lanka, revealed a seven-fold benefit of training within the first year, in terms of increased farmer income relative to training costs, and farmers continued to use IPM methods over the six year study period. Thus there is some evidence that the benefits of conserving natural enemies, following their disruption by pesticide use, may outweigh the costs. It is far more difficult to obtain estimates of the value of other measures to conserve or enhance the action of local natural enemies. Many measures which might enhance biological control by local natural enemies will have other kinds of economic impacts as well. Hence, measures like inter-cropping or zero-tillage, which may enhance natural enemy action, will affect the economics of production in other ways, making it difficult to isolate the specific value of the biological control element.

FUTURE TRENDS IN BIOLOGICAL CONTROL

What trends are likely in biological control over the next few decades? Conservation and augmentation of natural enemies will be considered separately from biological control introduction, because while all of these approaches will be affected by trends in pest management, changes in introduction will have an additional driver, namely trends in the occurrence of alien pest species.

Future pest management will have elements of natural enemy conservation, augmentation, pesticide use and other methods. The importance of conservation and augmentation in future will be strongly influenced by the future development of IPM, which treats these approaches as potentially important elements. It will also be influenced by the future of the chemical pest control. The emergence of genetically modified (GM) crop systems deserves consideration because it is spreading rapidly and replacing pesticide markets in the continued growth of the pest control industry. In the past 50 years, IPM has been adopted as a national crop protection policy by an increasing number of countries.

IPM has been interpreted differently by different countries and different approach have been taken, ranging from a commitment in USA to have 70% of crops under IPM, to actions by several European countries to reduce pesticide use by 50%. Critics have observed that commitments to IPM are not realised or that they focus on pesticide management and not natural enemy conservation. In a review of IPM, focused mainly on the USA, Kogan suggests that the great majority of IPM programmes simply use pest surveillance as a basis for managing pesticide applications, whilst incorporating natural enemy monitoring and methods such as augmentation, which characterise IPM as originally conceived, occurs in less than 10% of IPM programmes. IPM has provided an opportunity for the development of conservation methods and its spread will lead to a growth in conservation of natural enemies, but the rate of this growth will be very dependent on the extent to which IPM explicitly incorporates biological control elements.

Parallel to IPM trends in government policy, the food industry is playing a growing role in promoting elements of IPM. Largely driven by consumer concerns about pesticides, major food distributors and retailers are imposing IPM-like standards on their producers. This may take the form of certification schemes run by particular companies or by farmers associations. Organic production is growing and, while it continues to have a very small market share, it has considerable influence on public perspectives on pesticide use and sustainable agriculture.

More widespread integrated production schemes make use of natural enemies in their promotion – ladybirds are a common icon of integrated production labels on food - if not substantially as yet in their schemes. These changes are being driven quickly and on a global scale, in part as a result of the consolidation of the food industry around fewer, larger multinational retailers and the growth of global sourcing of agricultural production for these demanding markets. As with national IPM initiatives, the effect of this on biological control will depend on how explicitly integrated production requirements identify active natural enemy conservation and augmentation, rather than simply minimal and safe pesticide use. Within production systems, the future of augmentation is unclear. In the face of globalisation of agriculture, direct government support to agricultural production, for traditional reasons of food security, may decline and with it government-subsidised biological control augmentation programmes. Further, public sector augmentative production may be lost, particularly in the developing world in the face of cheap, generic pesticide alternatives.

For commercial augmentation, however, we have seen that there is a stable, niche market in certain crop systems like high value horticulture, that this market is growing and that there is much room for growth. There is some evidence that the European glasshouse biocontrol market is saturated, and possibly even declining with competition from new specific pesticides. However, only 5% of glass house production worldwide use biological control, so there is considerable scope for spread of this approach.

Guillon has estimated the 2003 global commercial market for augmentation at $425m (47% macrobial, 53% microbial), and the 2008 market at $573m, suggesting that slightly more of this 35% growth will be in microbials than macrobials. The principle constraint on the rate at which augmentation grows is presently the national regulatory systems that make difficult the registration of new projects, which is discussed further below.

The requirement and cost of pesticide registration and re-registration on minor crops is causing a dearth of effective pesticides in crop systems where augmentation and conservation have particular strength, which may prove a further stimulus to their growth. On the other hand, new chemical pesticides are being developed for a range of crop systems which are increasingly selective, in that they do not interfere with natural enemy action. While these products may complement natural enemy augmentation, they also have the potential to replace it, if more cost effective, as suggested above for European glasshouses.

The factors which influence the future of biological control conservation and augmentation are diverse and contrasting, making prediction difficult. On the basis of the spread of IPM and patterns of growth of commercial augmentation, it is reasonable to conclude that demand for biological control knowledge and technologies will grow steadily in future. This growth will be associated with the spread of IPM and augmentation in particular crop systems.

An increasing number of government and international agencies, farmer organisations and consumer groups are aware of IPM and its benefits, and it is likely that this will drive a conversion of systems to IPM and a preference for IPM in newly established crop systems. However, it is important to observe that this trend towards IPM will also be punctuated, as in the past, by crisis and reaction. In a comprehensive review of pest management and IPM, Way and van Emden predict that broad spectrum pesticides will continue as a major component of pest management systems well into the 21st century. Where IPM is the preferred policy at the national or local level, this will necessitate active conservation of natural enemies.

However, in many countries and production systems, IPM is poorly understood, Here, chemical pesticide use and misuse is likely to continue to generate the patterns of pest resurgence which have emerged in the 20th century. There is a strong history of such pest management crises initiating national and even regional IPM initiatives with long-lasting effects. While avoidable new pest control crises are regrettable, their potential to drive IPM development should be recognised and used. Such crises may be particularly apparent in developing countries undergoing agricultural intensification, such as those in sub-Saharan Africa and South and East Asia, and in new crop systems, such as those for bioenergy crops, where there is little experience of pest control. Finally, one of the least predictable drivers of future natural enemy conservation is the introduction of genetically modified (GM) crops. GM technologies have, so far, focused on reducing the impact of insect pests (Bt crops) and weeds (herbicide tolerant crops), with some limited activity in other areas, e.g. viral coat protein technologies to protect plants against viruses diseases. Resistance to more insects and viruses, and to fungal and bacterial diseases and nematodes are likely features of future transgenic crops.

All of these modifications potentially "compete" as pest management solutions with measures to conserve natural enemies of pests. They also compete with augmentation, particulary biopesticides, a particular example being the displacement of *Bacillus thuringiensis* from markets where Bt cotton has been adopted.

Will biotechnology, by producing more pest resistant crops, replace the need for biological control conservation and augmentation? It is difficult to see this occurring. Firstly, GM technologies will be targeted by companies and breeders against intractable pest problems on a crop that lack effective, alternative methods of control, as this is where future markets lie. Some areas where biological control is effective many not be easy targets for GM crop technologies. For instance, sucking insect pests (Hemiptera) have so far been particularly difficult to combat with GM technologies, whereas many of these are particularly amenable to biological control.

The lack of attention by industry and government to possible effects on natural enemies early in GM crop development generated very considerable concern about incompatibility with biological control and IPM. IOBC and other organisations have initiatied studies on non-target effects, and this has led to an international project to establish guidelines for the integration of GMOs with natural enemies and other components of IPM. While there are some specific areas of concern, the general impression so far is that GM crops do not pose the kinds of conflict with biological control that broad spectrum chemical pesticides did in their time and that reduction of broad spectrum pesticides on Bt crops allows natural enemy populations to grow. Recent studies in UK, where concern about the biodiversity impact of GM herbicide-resistant crops is particularly strong, have shown that in some GM crops increased herbicide use may decrease biodiversity, including presumably natural enemy diversity, although impact on biological control was not measured.

Considerably more research is needed on the way in which GM crops and biological control can be better integrated for farmers. Yang *et al.*, in an important experiment comparing conventional, IPM and Bt cotton production in China showed that farmers employing IPM generated better yields with Bt crops than those not doing so. The active conservation of natural enemies allowed reduction of pesticide use on minor pests, and thereby proved important to realising the full benefits of using GM crop technology.

Independent of any direct interaction, there is a real risk that a sector-wide pre-occupation with new GM crop solutions may simply displace public and private investment in biological control. At worst, we could see a repeat of the history of chemical control, where the domination of pest control technology by multi-national companies, with their powerful political influence, resulted in underinvestment in non-chemical approaches to pest management, which ended only when the dominant technology began to break down. For instance, the likely demise of *Bacillus thuringiensis* as a biopesticide in cotton systems employing Bt cotton may be the logical result of market forces, but will such changes affect the way in which government research funds, which are still so important to both biopesticide and GM crop research, are invested in future? To avoid this myopic, "silver bullet" approach to pest control repeating itself, there is a need for GM technology to be seen, with biological control, *as a component of IPM systems*. If this occurs, GM crop development may be biological control-neutral in future.

REFERENCES

Anne R. Leslie and Gerrit W. Cuperus (eds.) *Successful Implementation of Integrated Pest Management for Agricultural Crops*. Lewis Publishers, Boca Raton, Florida.

Bugg, R.L. and C. Waddington. 1994. " Managing cover crops to manage arthropod pests of orchards". *Agricultural Ecosystems & Environment*. Vol. 50. p. 11–28.

Hoitink, Harry A.J. and Peter C. Fahy. 1986. "Basis for the control of soilborne plant pathogens with composts". *Annual Review of Phytopathology*. Vol. 24. p. 93–114.

Long, R.F. 1999. "Use of bats to enhance insect pest control". p. 67–70. In: *Bring Farm Edges Back to Life*! Yolo County Resource Conservation District, Woodland, CA. 105 p.

Tjamos E.S. (ed.) *Biological Control of Diseases*. NATO ASI Series, Vol. 230, Plenum Press, New York, NY

9

Sustainable Livestock Production

Livestock development over the past three decades has been mainly directed towards satisfying the rapidly increasing demand for milk, meat and eggs in the urban centres. The required intensification of animal production was often limited by both an inadequate indigenous feed and animal resource base along with a fragile socio-economic capacity in most developing countries. In many cases, the political pressure exerted by the urban demand led to the importation of exotic stock and concentrate feeds. Technologies which proved successful in the socio-economic context of developed countries were applied. Technical institutions were also founded on similar rationale. Large amounts of public funds were used to support the modern production systems. The traditional livestock sector was also stimulated by the increasing prices for animal products but its growth was hampered by inadequate input supply and support services. The signs of unsustainability of such development began to surface in the past few years.

The sudden increase in grazing pressure and lack of capacity for conservation among the graziers led to the degradation of rangelands. Feed imports could not be sustained. Lower adaptability of imported stock did not allow rapid multiplication and crossbreeding raised the fears of erosion of the indigenous genetic diversity.

The pollution of water and air has been limited to the few areas of livestock concentration in intensive production units. The large-scale use of insecticides in the past for Tsetse control and sporadic pollution of waterways with acaricides may have had significant environmental consequences.

Apart from the manifestations of unsustainability, the overall increase in animal productivity in the past three development decades has been minimal. Increased external assistance *per se* did not have a lasting impact on the performance of the livestock sector in developing countries. Particularly, smallholder or rural livestock production did not pick up any significant momentum.

PATTERNS OF LIVESTOCK DEVELOPMENT

The high economic demand for milk and meat in developed countries has led to the establishment of capital intensive systems that require high animal productivity, high levels of concentrate feeding and highly mechanised or automated infrastructure. The production technologies developed over the years have successfully met these requirements. The problems currently faced by these systems arise basically from economic pressures and revolve around capital costs, input costs and subsidies.

There are also environmental problems associated with high concentration of stock, animal wastes and high levels of fertiliser use for pastures. Feed availability is not a constraint in feed deficit areas thanks to the well-established trade in feedstuffs. Large quantities of feed are imported from developing countries.

In 1988, some 3.7 million tons of cereal brans and oilseed cakes were imported in the developed countries from the feed-deficit developing countries. The prevailing production systems in the developing countries are a world apart.

Large numbers of ruminant livestock graze common grassland especially in Africa. This low-input system had, in the past, successfully met a large part of the demand for meat in urban centres and catered for the local demand for milk and meat. For the past 2–3 decades the system has not been able to meet the increased demand arising from population explosion, increased income and rapid urbanisation.

Livestock numbers on common grazing areas have increased way beyond the carrying capacity. The lack of economic capacity among the graziers to undertake range improvement resulted in degradation of most of these grasslands. After an increase in numbers and offtake during the past 2–3 decades, the present output is declining. The degraded common grasslands cannot be rehabilitated without massive investment and the politically difficult land tenure policy.

A sizeable proportion of the third world livestock is also kept in arable farming areas and is integrated in the system to provide draught power, fibre and dung as well as marketable quantities of milk, meat and eggs. A few holdings in these areas are large enough to warrant increasing the producing ability of the stock. On small farms, the large ruminants are kept primarily for work; milk and meat are by-products.

Small ruminants are primarily kept for meat; milk is a by-product. Crop residues are the main feed resources although a small part of the land is devoted to fodder for work animals. Poultry and pigs basically scavenge the farm and household wastes. Supplementary feeding depends on the market for produce.

In recent years, the increased price of milk and meat and improved marketing opportunities for the small farmers have, in some cases, resulted in increased fodder cultivation, better utilisation of crop residues and by-products, supplementary feeding and animal husbandry geared towards marketable milk, meat and egg production. However, except for poultry, these improvements have not taken root to a significant extent. Spurred by high demand for milk, meat and eggs in urban centres, heavy investments were made in dairy farms, ranches and feedlots following the developed country model. This was facilitated during the mid-seventies to mid-eighties by growth in external loans, foreign exchange earnings and availability of capital at low interest rates.

Figure 1. Cows in field

In most countries which encouraged this model the available foreign exchange was used to import high yielding stock and feedstuffs. However, the contribution of these enterprises to total milk and meat supply has been minimal and at high cost. With increased capital and foreign exchange shortages in recent years, the specialised livestock farming enterprises and feed industries are trying to re-orient their operations towards greater reliance on locally available feedstuffs and animal resources. There is a growing realisation of the value of multipurpose local stock and their capacity to utilise fibrous feedstuffs. A major part of milk and meat in developing countries is produced by the resource-poor small farmers. Smallholders contribute by far the largest part of the labour force. To improve production and income, techno-economic changes must be brought in the small farming system. Availability of sufficient feeds of adequate quality is the basic constraint in the system.

Possibilities of providing external feed inputs are limited. Due to a lack of sufficient attention to the smallholder, appropriate technologies to improve performance of locally available animal and feed resources within the rural system have not developed. Institutional capacities in developing countries have also not been built up from this viewpoint. Smallholder dairy development, for example, has succeeded only when appropriate production technology was supported by an integrated dairy development programme incorporating milk collection, input supply and favourable price policy.

TECHNOLOGY REQUIREMENTS

The objective of livestock development at the national level has been to attain as much self-sufficiency as possible to satisfy mainly the urban demand. At the farm level, the objective is to increase income and utilise family labour year-round. The role of animal production technology in developing countries revolves around finding labour-intensive procedures which maximise production with low-cost inputs. Feeds comprise by far the largest component of input cost.

The efficiency of feed utilisation in terms of herd or flock output is the primary consideration in technology development. Other elements of the development of appropriate and affordable technologies that ensure sustainability are: conservation and improvement of resource base; minimisation of wastes and environmental degradation; and recycling of wastes for animal feed or biogas. Direct transfer or transplantation of technology from developed to developing countries has rarely been successful and sustainable in achieving the above objectives. However, in the case of a few specialised enterprises where such a transfer has been successful, the longer term consequences have been dependent upon imported feeds and exotic animals in order to take maximum advantage of the transferred systems.

The side effects of imported technologies has been the neglect of indigenous livestock and feed resources. A sustained growth of indigenous production of milk and meat is dependent upon the introduction of new technologies that would be adopted by producers, especially, the smallholders.

A systems approch is indispensable for the development of such technologies. This approach should incorporate the rigour of the scientific method as well as the human element of involving the producer of the farmer at all stages of development. There is a pressing need to utilise the effective tools and techniques that are available to pursue a farming systems research and extension strategy.

The farmer must be a key partner in every effort in the improvisation and introduction of the required technologies. Multi-disciplinary effort and effective linking of research, extension and training are important requisites for the systems approach. The process of increasing animal productivity must also be effectively linked with other developmental elements such as market considerations, produce organisation, incentives and policy-making. Vigorous research-extension efforts have proved to be the basic requirements for improving animal production systems.

During the past three decades, most developing countries have been able to establish institutional infrastructure for livestock development, e.g. research centres, extension services, veterinary laboratories, disease control services and educational institutions at various levels. The technical performance of this infrastructure is variable from country to country and from institution to institution. On the whole, the institutional impact on livestock production has been open to question.

The development concern these days is not so much about the capacities in terms of physical infrastructures or size of trained manpower but about the usefulness of this capacity in improving farm output. The underlying constraints emerge from the prevailing traditional, neo-social, economic and political environment. Self-reliant livestock production in developing countries requires a strategy to optimise production from available feed resources through an integrated technology which employs multi-purpose crops, multi-purpose animals and recycling of residues and by-products.

The identification of needs and a careful study of feed and animal resources are essential first steps. Resources must be examined in the context of various agro-climatic zones and fodder crops which might be grown. Ruminant production systems must then be matched with the resources in a way that aim for economic rather than biological maximisation. To introduce new technologies it is important to start with on-farm improvements.

The utilisation of locally available feed resources must be maximised to reduce or eliminate the importation of concentrate feeds. Livestock development efforts in the past laid primary emphasis on rapid genetic improvement arguing that improvements in feeding will be ineffective when animals with low genetic potential are raised. In recent years there is a growing consciousness to balance the rate of genetic improvement with improvements in feed availability and management. There is also an increased realisation of the potential of indigenous cattle, buffaloes, sheep and goats as multi-purpose animals suited to sustainable production systems and as efficient convertors of locally available feed resources. Innovated breeding and management procedures have not been effectively implemented to improve reproduction and to increase the milk, meat and work outputs of the indigenous stock.

ISSUES IN SUSTAINABLE ANIMAL PRODUCTION

Technological Problems

Developing technologies for sustainable animal production has always been the objective of animal scientists even this was not expressly stated. This concern had been well taken into account regarding the environment in which the technologies were developed. However, the use, rather than misuse or misplacement, of a proven technology could be an issue in the context of sustainability in different environments.

The important issue in this regard is the development of appropriate and affordable livestock technologies suited to specific agroclimatic zones in developing countries. To satisfy the criteria for sustainability, these technologies should support agricultural development which "conserves land, water, plant and animal genetic resources, and is environmentally non-degrading, economically viable and socially acceptable". This would mean further research and development effort for the expansion of the feed base, conservation of animal genetic resources, recycling of wastes and efficient feeding systems. Methodology to evolve sustainable production systems requires a multi-disciplinary approach and development of appropriate indicators of medium-term or long-term sustainability. These indicators should be practical enough to be used in project formulation, monitoring and evaluation.

Institutional Issue

The institutional framework for improving production systems needs to be structured, or restructured, to ensure multi-disciplinary collaboration and cost-effectiveness. Participation and cost-sharing by the farmers' organisations may be necessary in most cases. Present institutional structures for research, extension and veterinary services need evaluation.

Import/Export Issues

Past experience has highlighted the unsustainability of livestock production and feed imports in most developing countries. The impact of imports on the existing production systems needs to be fully understood before allowing the import of milk, meat, feeds or live animals. Similarly, the export of agro-industrial by-products from feed-deficit countries disrupts livestock development efforts.

Price policies for milk, meat, feed or eggs should be geared towards creation of an economic environment in which it would be profitable to conserve and utilise local feed and animal resources for production. These policies should be devised and implemented for the benefit of the producer.

Degradation

Common grazing lands are most vulnerable to degradation. The basic issue in rangeland rehabilitation is land tenure policy which would encourage range improvement and proper grazing management. Access to the rangelands by the nomads and other poor segments of the population is also an issue.

Conservation Problems

The relevance of the conservation and genetic improvement of indigenous animal genetic resources is well recognised. The important issues concern the cost and the cost-effectiveness of methods to be employed. It has often been mentioned that these efforts should be supported by public funds and external assistance. However, this point of view should not divert the attention from efforts to develop privately-supported and cost-effective methods for conservation.

STRATEGIES FOR SUSTAINABLE LIVESTOCK DEVELOPMENT

The sustainable development of animal agriculture, across the many different ecoregions or agroecological zones that are found in the developing countries, poses many fundamental challenges; challenges to the primary users of livestock, to their extension, research and support service agents (private- and state-sponsored), to local and regional development authorities, to investment banks, government policy makers and their institutional organs and, in the final analysis, to the consumer or user of animal output.

These challenges have been confronted by mankind with varying degrees of intensity since man first captured and domesticated wild animals for use. However, it is only in the past 200 years that what may be described as an industrial approach to animal agriculture became manifest. Over the past 40 years, the United Nations, investment agencies such as the World Bank, and many bilateral and non-governmental organisations, all in consort with national governments, have launched development programmes to promote or effect the advancement of livestock production throughout the developing world. There have been many successes, particularly in pig and poultry production but, unfortunately, many failures also.

Extensive reviews of livestock investment projects conducted by the World Bank and the Asian Development Bank paint a gloomy picture of the success of development investment in livestock, particularly in the smallholder sector which farms the vast majority of all livestock in developing countries. This lack of success/impact, in turn, has led to a significant reduction in the amount of investment funds directed to livestock production.

Added to this, there is increasing interest in all issues related to the sustainability of agricultural development programmes, including livestock production, not only those factors that may directly effect the environment (e.g. Co_2, methane), but also a more thorough scrutiny of development inputs and technologies in terms of longterm practical application and impact.

Determinants of Sustainability

Effective development planning entails a comprehensive understanding of man's need to advance development, a thorough analysis of the technical, economic and social implications of proposed interventions and finally, an assessment of potential environmental impact (long-and short-term) that may influence local, regional or global conditions.

The development audiences, i.e., the individual farmer, the extended family or local community and the regional and/or national governments, often have very different perceptions of what development means. Technical interventions may be viable economically, but may not be self-supporting and may not be socially acceptable if, for example, they result in increased or conflicting work routines within the farm family.

Finally, whereas few farmers in developing countries are acutely concious of the global environmental impact of their actions, they certainly are aware of their local habitats and, generally speaking, respect traditional practices and taboos that condition their farming patterns.

Production Objectives

The development of intensive livestock production in industrial countries over the past 30 years has become synonymous with single purpose breed specialisation; consequently, Holstein Friesians for milk production and continental beef breeds such as the Charolais, automatically spring to the minds of livestock development planners throughout industrialised Europe, America and Australasia.

Unfortunately, some of these people often export their 'vertical thinking' into livestock development planning in many developing countries; countries to which specialised, single-purpose breeds are not suited and can rarely adapt to the harsh local environments. Market demand, labour and capital costs, together with land availability, have dictated specialised single-purpose animal production systems in the industrial countries; production efficiency is ultimately measured in terms of shelf price and continuity of a standardised, high quality, animal food product. However, interesting contrasts are found within industrial countries, if, for example, sheep management systems in a land-rich, low population country such as New Zealand, are contrasted with a high population European country such as Britain.

In New Zealand, sheep breeding goals are focused on relatively low producing (lambing percentage, 90%-100%), easy-care sheep, where one man can manage approximately 4,000 ewes. In contrast, British lambing percentages in the smaller flocks (200-400 ewes) are targeted at levels approaching 200%. However, the contrasts in animal breeding objectives between industrial and developing countries are much more stark.

Figure 2. Goats

Several livestock development projects have failed to attract committed farmer participation simply because they were targeted on single-purpose milk or meat output and ignored the farmers primary interest in livestock as providers of animal traction or manure. Further, the bioenergetic efficiency of multipurpose livestock production is often overlooked by livestock planners and, consequently, misplaced development objectives are often pursued.

Examples of successful screening have been reported in Ireland and Britain. In Ireland, the national sheep flock was screened for ewes displaying exceptional litter size, i.e., 3, 4 or more lambs/ewe. The open nucleus flock which was then established had an average litter size of 2.3 lambs/ewe as compared to 1.3 in the national flock. Analysis of litter size distribution and repeatability estimates in this flock suggests the presence of a major gene or block of genes controlling ovulation rate. The Javanese Thin Tailed sheep (JTT) is also showing clear evidence of the existence of a major gene controlling ovulation rate. This latter example is cited as evidence that major genes may and have been found in livestock populations living in harsh environmental conditions.

Screening of Awassi Ewes for Milk Yield: FAO initiated a preliminary programme to test the efficacy of genetic screening (GS) in conjunction with the establishment of Open

Nucleus Breeding Scheme (GS/ONBS), as stage one in the genetic improvement of livestock breeds in developing countries.

Genetic Improvement Capacity

Very often development planners recommend a breed development programme, which, while technically justified, is without adequate emphasis on and evaluation of the capacity and commitment of the different 'actors' in the breed development chain.

Two issues are paramount here: that livestock breed development is a long-term problem requiring, at a minimum, a 20-year commitment; and that breed improvement cannot have any meaningful relevance if it does not compensate the farmer and the other participants in the breed improvement process over a sustainable timeframe and, to this end, research personnel, extension staff and support services must play strong and consistent roles to support effective breed development and farmer participation. Further, long-term government policy and its impact on the development environment will also condition breed improvement possibilities. Consistency and coherence across all the areas in the breed improvement process, broadly determine genetic improvement capacity; all links in the chain are important to the final outcome.

Integrated Crop-Livestock Systems

The livestock feed resource base within any given land-use farming pattern determines animal productivity; conversely, the land-use pattern sets bounds on the role(s) of livestock within the farming system. The challenge to livestock development planning is to strike acceptable and sustainable balances between these sometimes conflicting interests.

Land use

The still rapidly expanding human population in most of the developing world, which is projected to increase from 4 to 7 billion by 2025, continues to focus attention on land use, not only to meet the populations' staple food but also their living space requirements. The trends to urbanisation across sub-Saharan Africa, West Asia/North Africa, Asia and Latin America suggest that 54, 75, 56 and 84%, respectively, of the populations on these continents will live in cities by the year 2025. Both of these fundamental demographic forces demand the attention of livestock planners now and towards the future. They dictate and emphasis on land-use efficiency in which the synergisms of soil, plant and animal interactions must be fully exploited.

Coupled to this, concerns to conserve the natural resource base and to limit the emissions of carbon dioxide, methane and other environmentally damaging impacts, will further add to the complexities of livestock development planning. Bioenergetic efficiency must become the keystone of livestock farming systems in the future. Inevitably, this means integrated crop-livestock systems throughout the tropics and subtropics. Certainly, in the low rainfall (<200 mm/year) rangelands, or in the high altitude alpine pastoral systems, crop livestock interactions will not be strongly evident, but in these environments avoidance of rangeland degradation through overstocking will be of paramount importance. Quite distinct but ultimately compatible short and long-term strategies are required.

Traditional feed resources: In the short term, the development of feed resources to raise ruminant livestock productivity throughout most of the developing world must be based on better use of traditional feeds. Generalisations are difficult to make across the diversity of production systems and their very different feed crops. However, the first step in all systems must be to tackle critical nutritional deficiencies, be they mineral-, protein- or energy-limiting. In extremely difficult conditions, this may simply mean strategic supplementation of the diets of selected poorly thriving animals in order to reduce mortality.

At another stage it will mean improving feed utilisation through: the treatment of by-product feeds (ammonia, urea); diet supplementation with balanced highenergy feeds, (urea/molasses blocks, etc.), coupled with the feeding of by-pass protein. The underlying strategy being to build on the existing system and introduce simple and practical technologies to suit local conditions. Parallel to improvement in the feeding regime, due attention must be given to the major prevailing animal disease challenges. In the first instance, this will usually call for parasite control (internal and external parasites) coupled with vaccination against specific diseases as necessary.

New Feed Technologies: In the longer term, adequately supported by ongoing and future research, greater emphasis will need to be focused on the development and utilisation of high biomass feed resources (e.g. sugar cane, cassava) for monogastric and ruminant livestock, coupled with better exploitation of high protein forage trees. Expansion of livestock production in the tropics in the future will demand technologies that enable the ruminant to better utilise high fibre feeds, e.g. rumen flora manipulation to improve digestion of cellulose. Coupled with these changes, livestock planners in the future are likely to have to place much greater emphasis on multipurpose farming systems in which integrated livestock, food and tree crop production maximises photosynthetic capture to produce food (plant and animal), fibre, energy and fuelwood while maintaining soil fertility and overall sustainability within the system.

Examples of such integrated systems are already being developed to exploit the high biomass potential of the tropics. In other areas of the tropics, the more extensive exploitation of integrated rice, fish, asolla and livestock production is being researched in order to develop bioenergetically-efficient production systems to meet food needs in land-scarce countries such as China and South Asia.

Development Force

Technical strategies for the advancement of animal agriculture constitute just one vector in the overall matrix of effects that determine the ultimate impact and sustainability of development effort. To place them in their proper context, it is useful to represent the development planning process as an algebraic function of the major determinant vectors.

For sustainable livestock development, these vectors will encompass: market demand for animal product or use, a_i; market/animal use channels and pricing arrangements, b_i; land and animal resource potentials and interactions, c_i; practical technical interventions and transfer impacts, d_i; natural resource sustainability and environmental impacts, e_i; and the major human components of the development process, f_i. This last category ranges from farmer initiative and welfare to the efficiencies of the major government and private support service personnel involved in research, training, extension, market regulation, provision of capital and the formulation and implementation of government policy; these parameters of human endeavour might be conveniently termed the 'development force'.

The challenge, inherent in the development of strategies for the sustainable advancement of animal agriculture, is to fully comprehend the direct (additive) and interactive (non-additive) effects of these different determinant vectors. They are represented in algebraic terms in the following equation, although it is fully understood that it is not possible to emperically formulate such an equation. Modeling and maximum likelihood iterative procedures can help to establish the relative importance (B) of each determinant vector on sustainable livestock development (SLD).

Approached in this way, livestock development planning requires an interactive approach in which a number of development scenarios are identified; the interpretation of the outcomes will demand coordinated interdisciplinary analyses involving economists, sociologists, production scientists (land and animal) and environmental impact physics and chemistry.

Farmer Welfare

Many development programmes are based on the 'perceived' needs of the target farmers; the word 'perceived' is generally used by planners to indicate a full understanding of what the farmer needs! Farmer needs, like those of any other sector, are a function of tradition, education, and day-to-day survival demands normally influenced by the aspirations of the individuals, the extended family and the local community.

Increasingly, farming systems research and extension (FSR/E) approaches are attempting to put real perceptions on these needs; the livestock development planner must follow suit, essentially by 'standing in the farmers shoes' and perceiving development needs and objectives from that position.

Concerns for such matters as 'internal rate of return to investment', high output production systems and impacts of new technology will have to be transposed on such vital issues as investment risk, demand on work routines and family welfare since these are influenced by local tradition, culture and status in the local community. The development planning task is to marry these very different perceptions in an acceptable and phased development programme. The roles of the extension and other support services (e.g. credit) have a very important role to play in this reorientation/transition process.

Extension/Research Service Efficiency

Most developing countries, particularly in Africa, dramatically expanded their research and extension services over the past twenty years, basically based on World Bank investment loans. Yet it is difficult to single out many countries that can be deemed to have effective research and extension services to support the development of livestock production.

Several International Service for National Agricultural Research (ISNAR) and other studies have identified and articulated many of the reasons why these services have not been successful in supporting change and sustainable advancement. Only two of the many reported constraints on research and extension services in developing countries, discussed in these studies, will be mentioned here; staff motivation and research-extension-support service links. These are often recognised, but are not always given adequate attention by development planners simply because they fall within the realm of national government policy and management capacity.

Staff Motivation

Staff motivation is certainly influenced by training, it can be nurtured or dissipated by management but, in the final analysis, it is heavily dependent on financial reward to the individual. Many research-, extension- or development-oriented projects have experienced great difficulty in harnessing effective commitment from national counterpart staff simply because the staff in question, albeit formally working full time in government service, find it necessary to have a second or sometime a third additional job/activity to earn enough money to support their families. It is unreasonable to accept that development objectives can be realised in such circumstances, although they may be very appropriate in every other way.

Support service efficacy is also heavily dependent on effective professional interaction between public service staff employed by different government departments or agencies. Sadly, it is not uncommon to find situations where apathy, jealousy and sometimes hostility discourage and limit the interactions, exchange of information and cooperation between the different sectors of government support services that are so essential to effective guidance and development of their farmer audiences. The sustainable development of animal agriculture, with its many complex interactions demands that these human initiative issues are fully considered in consort with the other elements in the sustainable development matrix.

REFERENCES

Bommer, D.F.R. and A.W. Qureshi, 1988. *Livestock development projects: successes and failures.* Proceedings VI World Conference on Animal Production, pp. 198–212. Helsinki.

Eberly, Eric, and Gordon Groover. 2001-2002 *Virginia Farm Business Management Crop and Livestock Budgets.* Virginia Cooperative Extension. Publication number 446-047. p.7

FAO. 1989. *The State of Food and Agriculture. Part III. Sustainable Development and Natural Resource Management.* Rome.

Preston, T.R. and R. Sansoucy, 1987. *Matching Livestock Systems with Available Feed Resources.* FAO Animal Production & Health Paper No. 63 pp. 32–49, Rome.

Winrock International 1992. *Animal Agriculture on Developing Countries: Technology Dimensions.* Morilton, Arkansas.

World Bank. 1985. *The Small Holder Dimension of Livestock Development: A review of Bank Experience,* Report No. 5979, Operations Evaluation Department, 1985.

10

Sustainable Aquaculture Development

Aquaculture is the farming of aquatic organisms, including fish, molluscs, crustaceans and aquatic plants. Aquaculture has been the world's fastest growing food production system for the past decade. Aquaculture production increased from 7,4 million tonnes in 1980 and 16,8 million tonnes in 1990 to more than 42 million tonnes in 1999, valued at over US$ 53 thousand million. The sector's production is growing at an average rate of more than 10% per year, as compared with a growth of about 3% for terrestrial livestock meat production, and 1,5% for capture fisheries production. The contribution of aquaculture to world food fish landings has more than doubled since 1984. In 1997, over 30% of food fish consumed by humans, from a total average per caput food fish supply of 16.1 kg, was provided by aquaculture. Global projections for future supplies from aquaculture production include, for example, 47 million tonnes for the year 2010.

Asian aquaculture farmers continue to contribute about 90% of the world's aquaculture production, and in 1999 more than 82% of total aquaculture yield was produced in low-income food-deficit countries (LIFDCs). The growth rate of the aquaculture sector in LIFDCs between 1984 and 1995 was six times faster than that for non-LIFDCs.The vast majority of finfish produced by aquaculture is based on extensive and semi-intensive freshwater culture systems producing predominantly Chinese and Indian carps, and contributing more than 44% of global total aquaculture production by weight in 1999. In contrast, marine and brackishwater aquaculture systems employed in coastal areas inf 1999 yielded 23,4 million tonnes, valued at US$ 30,3 thousand million, representing 55% of total volume and 56% of total value of global aquaculture production.

Coastal aquaculture is dominated by production of aquatic plants (seaweed) and molluscs. For 1999, their shares of total coastal aquaculture production in terms of quantity and value are 40% and 19% (seaweed) and 43% and 30% (molluscs). The production share of crustaceans (6%) and finfish (11%) is comparatively low, but their relative contributions to the value of total coastal aquaculture production is significant, i.e. 24% (crustaceans) and 27% (finfish). A wide range of very diverse coastal aquaculture systems has been developed in Asia, Europe, and the Americas, operating with different intensities and scales of production. The potential for additional growth and future expansion of coastal aquaculture is being recognised by many government authorities, private sector (investors, aquaculturists and ancillary activities), financial institutions, such as development banks, as well as aid agencies, at national and international levels.

COSTS AND BENEFITS OF COASTAL AQUACULTURE DEVELOPMENT

Planning for sustainable development and improved natural resource management implies a thorough examination of different development options in terms of their financial, economic, social, and environmental costs and benefits, and the distribution of these costs and benefits through time and space, and between different groups in society. This implies some form of valuation—either qualitative or quantitative. Despite the rapid growth of aquaculture and the growing awareness of environmental issues, few studies have been made which address these issues objectively. In many cases the debate has polarised between those who emphasise the economic benefits, and those who emphasise negative environmental impact. The debate has also tended to generalise from specific examples, although the sector is enormously diverse. For example, four major species groups are farmed in coastal areas, including seaweed, molluscs, crustaceans and finfish, with a range of significant differences within and between each of these groups.

Planning and Management

Experience has shown repeatedly that without some form of intervention, short term financial perspectives will tend to dominate development decisions to the detriment of environmental and social objectives. In the case of coastal aquaculture, and indeed many activities in the coastal zone, there is a strong case for such interventions to be planned and strategic, rather than reactive and uncoordinated. The problems associated with coastal aquaculture development may be grouped into three broad categories as follows:

- unsuccessful development, where the potential for development is not realised, especially among the poorer sectors of society;
- the vulnerability of aquaculture to poor water quality and aquatic pollution, caused by industrial, domestic, agricultural and aquacultural (i.e. its own) wastes;
- over-rapid development, where the undoubted successes of the sector have been tarnished by environmental and social problems, disease, and in some cases, marketing problems.

Investors have responded to these problems with more rigorous project appraisal: financial and economic analysis, and in some cases cost benefit analysis. Governments have responded with specific regulations relating to farm operation, and/or with more rigorous requirements for social and environmental impact assessment (EIA). The market itself is increasingly demanding sustainably produced goods, at least in western countries.

These responses have significant weaknesses. They arise mainly from the small scale and incremental nature of most aquaculture development. While individual developments may have no significant impact on the environment or society, a large number of developments, however small, may have significant impacts on the wider social and economic environment, and on each other. Farm drainage in western countries, and shrimp farming in some regions of Asia are classic examples of this problem. Project or enterprise level approaches cannot deal with this problem, and the market is likely to respond only once damage is done. Furthermore, EIA and economic/ financial studies tend to be undertaken by different specialists, ignoring the close links between the two, and commonly presenting contradictory conclusions.

Nor can these approaches facilitate or promote aquaculture development in those areas to which it is most suited. This is a particular problem with aquaculture, because site requirements are frequently much more demanding than those for other activities. Inadequate attention on the part of new entrants to site selection is a major cause of failure in aquaculture development, and commonly exacerbates environmental impacts.

In practice, the problems and opportunities associated with coastal aquaculture development can only be addressed or realised through:

- improvements in siting, design, technology, and management at the farm level (requiring a set of incentives and constraints to promote these changes at the sector level);
- better location and spatial distribution of the sector as a whole (implying some form of zoning);
- better water supply for the sector as a whole;
- better fish health management, including disease and stock control at individual farm and sector levels;
- improved communication and information exchange;
- improved access to markets and trade opportunities; and
- more equitable distribution of the benefits derived from coastal aquaculture development.

This implies strategic intervention by government and producer associations or industry organisations to allocate and use resources more equitably and efficiently in both time and space—in other words, more effective and integrated planning and management of the sector.

Scope of Coastal Management

Coastal management implies something broader than addressing the development or resource issues associated with one particular activity or sector in the coastal zone. Olsen *et al.* have proposed the following typology of coastal management:

- *Enhanced Sectoral Management (ESM).* Focus on a single sector or topic but explicitly accounts for impacts and interdependencies with other sectors, ecosystem functions, and institutional capacity.
- *Coastal Zone Management (CZM).* Multi-sectoral planning and regulation focused upon the characteristics and needs of narrow, geographically delineated, stretches of coastline.
- *Integrated Coastal Management (ICM).* Expands the cross sectoral feature of CZM to consideration of the closely coupled ecosystem processes within coastal watersheds and oceans

In practice there is a broad range, or continuum, of coastal management initiatives relevant to aquaculture which are more or less integrated in terms of geographical scope, horizontal (sectoral) integration and vertical (policy) integration. CZM and ICM type initiatives in particular overlap significantly in practice, and are difficult to assign to these sub-categories. For practical purposes they are therefore grouped together in the discussion below.

Enhanced Sectoral Management (ESM)

Initiatives which seek to enhance the sectoral management of aquaculture and aquaculture development are widespread in developed countries. Interest in these approaches has been generated because of heightened awareness of sustainability issues in general, and those related specifically to aquaculture. Enhanced sectoral management tends to be conservative in nature, the initiatives usually arising within existing institutions, and based on existing responsibilities and powers.

These initiatives are diverse. They include the use of environmental impact assessment (EIA) at sector or farm level, and/or a package of tools and incentives to promote better siting or more sustainable practices. In some cases assessments of environmental capacity and its relation to the quantity or location of aquaculture production have been undertaken. Other initiatives focus on the identification of suitable sites or zones for aquaculture development, facilitated through the use of GIS or remote sensing. In Tasmania, a more comprehensive approach is now in place that requires the development of local marine farming development plans. Initiatives in this category have been funded by government at all levels, by development banks or aid agencies, and by private companies. However, the lead institution is generally that with traditional responsibility for fisheries and aquaculture.

Strengths

- avoids the risks of more radical approaches to institutional change (e.g. confusion over powers and responsibilities; lack of institutional capacity);
- builds on and enhances existing knowledge and skills;
- allows for relatively rapid assessment and research, followed by implementation of improved planning and management of aquaculture development—*if* appropriate institutional powers and capacity already exist.

Weaknesses

- sometimes "top down" and driven by technological priorities and considerations;
- the values and concerns of other stakeholders may be inadequately understood or taken into account;
- where the lead agency also has responsibility for aquaculture development, the initiative may be biased in favour of development;
- assessment, evaluation and monitoring of the effectiveness of interventions may not be objective or effective, because of a lack of external perspective;
- may ignore rather than resolve conflict;
- may end up as paper research and project style exercises, if planning and regulatory powers and institutional capacity are lacking;
- may partially duplicate the efforts of other sectoral agencies also involved in enhanced sectoral planning;
- may be more costly in the long term, and less consistent and effective than more integrated approaches.

Coastal Zone and Integrated Coastal Management

Coastal Zone Management, Coastal Area Management, and Integrated Coastal Management have been widely proposed as more comprehensive approaches to coastal management which address the limitations and difficulties associated with sectoral and enhanced sectoral approaches, particularly in relation to aquaculture. Coastal zone management implies multi-sectoral planning and regulation, and therefore some form of co-ordinating body or authority to assess and balance the various sectoral interests. ICM also implies mechanisms for addressing trans-boundary issues.

CZM and ICM initiatives have varied enormously in terms of specific objectives, overall approach, geographical and sectoral scope, initiating or implementing institutions; and in terms of the influence they have on decision making and resource use in coastal areas. They have arisen from academic or political initiatives, aid funded projects, or directly in response to an environmental problem or development need. Nonetheless, most ICM initiatives have certain key features in common. The goals or objectives usually include reference to one or more of the following:

- the optimal allocation of resources to competing activities or functions;
- the resolution or minimisation of conflict;
- the minimisation of environmental impact, and the conservation of natural resources.

In some cases ICM may also have more strictly social and political objectives, such as quality of life; the more equitable distribution of derived economic benefits; social and inter-generational equity; and poverty alleviation.

There have been many reviews and guidelines related to ICM published in recent years. Although there is a broad consensus as to the main components of ICM, emphasis and details vary widely. Cicin-Sain *et al.* compared coastal management guidelines developed by five different international entities. Based on their comparisons, the authors developed a "consensus set of ICM guidelines".

Aquaculture and integrated coastal management

In many ways aquaculture is a classic example of why ICM is needed:

- coastal aquaculture commonly straddles the boundary between land and sea;
- resource (land, water, and their products) ownership or rights allocation, and related administration, is often complex or ambiguous in prime aquaculture locations;
- aquaculture may be seriously affected by water quality and habitat degradation caused by other activities;
- aquaculture itself may affect environmental quality and the interests of other users through conversion of natural habitat, through pollution of recipient waters with nutrients, organic substances, and potentially toxic (hazardous) chemicals, and through the spread of disease;
- poorly sited or planned aquaculture may result in negative feed-back and self pollution.

Unfortunately, there are few clear examples of the successful integration of aquaculture into comprehensive ICM. It is arguable that this is because there have been very few genuine ICM initiatives, where aquaculture has been assessed alongside the full range of existing or potential activities in the coastal zone using consistent and rational assessment criteria, agreed across a range of interests and agencies. However, to do this thoroughly takes time, and this poses a dilemma in many developing country situations where aquaculture is developing very rapidly. The case of Ecuador, where population pressure, industrial development and shrimp farming have had significant negative impacts on estuarine resources throughout a period in which a long term ICM project was underway, is particularly notable. Shrimp farming has also recently developed uncontrollably in Sri Lanka, with adverse environmental consequences and self-pollution, despite a strong ICM awareness, and a variety of ICM initiatives in place.

In other countries, where existing institutional and planning structures are favourable, and where the development pressures are less extreme, ICM may be both desirable and feasible. An example is the case of New Zealand.

Strengths

- the values and concerns of the full range of stakeholders are specifically taken into account;
- relevant institutions are encouraged to communicate, co-ordinate and co-operate;
- a broader base of information and opinion is available to decision makers;
- less technically driven than sectoral approaches;
- more "bottom up" than "top down" (if correctly implemented);
- potential development activities are assessed objectively, using a broad range of criteria, against all other possible resource uses, not just those from the same sector;
- the resolution of conflict and balancing of interests is usually a specific objective;
- should result in consistency of policy and legislation between different levels and sectors of government.

Weaknesses

- may make inadequate use of existing institutional memory and skills: new institutions need to learn—they may repeat past mistakes or re-invent old solutions;
- since it requires institutional change, it may carry institutional risks (e.g. confusion over powers and responsibilities, and lack of institutional capacity);
- takes much time, effort and cost;

- may generate a wealth of assessment and research data with little consensus on how it should be used (as for enhanced sectoral approaches);
- may (at least in the early stages) exacerbate conflict by addressing and highlighting differing values and perspectives on resource use in the coastal zone.

GUIDING PRINCIPLES

Notwithstanding the diversity of experience and approaches, it is possible to present a set of core principles, which should guide as far as possible the development of any aquaculture planning and management initiative, whether it be a local initiative, an enhanced sectoral initiative, or more comprehensive ICM. They may be summarised as follows:

1. the Rio principles: sustainable development; the precautionary approach; the polluter pays principle;
2. integration or co-ordination: with other sector activities or plans; with national sector plans; with ICM where such initiatives exist;
3. wide ranging public involvement;
4. thorough assessment of costs and benefits (financial, economic, social, environmental) of aquaculture in a specific area (e.g. estuarine or lagoon system) and comparative assessment of costs and benefits of aquaculture relative to other resource uses;
5. some assessment of environmental capacity;
6. use of incentives rather than regulation where possible;
7. emphasis on the control of effects, rather than the scale of activity;
8. evaluation, iteration and adaptation; and
9. effective institutions and representative organisations.

Adherence to Rio Principles

Agenda 21 generated at the Rio (Earth) Summit emphasises sustainable development, and in particular its core value or principle of inter-generational equity. It also states a commitment to the precautionary approach, and the polluter pays principle.

Inter-generational equity cannot be defined simply in practice, but needs to be incorporated as a recurrent theme in the assessments, discussions and stakeholder exchanges related to different development and resource use issues. Aquaculture, like other development activities, may change the balance and distribution of different resources or *capital*. This capital includes natural capital; human capital; physical plant (equipment, machinery, buildings) and infrastructure; financial capital; and other forms of capital valued by particular societies. These changes must be assessed to ensure that the sum total of this capital, or specific vital components, are sustained or increased in the long term, and available for future generations.

The precautionary approach implies that we should more carefully plan and rigorously evaluate developments that have uncertain implications for the environment. Under conditions of great uncertainty developments may have to be delayed or halted. This principle is controversial, since its widespread application could slow or halt much development activity. Furthermore, the most successful development has often been associated with substantial financial and environmental risk and uncertainty. This principle should therefore be applied with care, taking full account of both the magnitude and likelihood of adverse environmental impacts. This implies some form of risk assessment.

The polluter pays principle is now widely agreed, and is a central tenet of much environmental policy. It is subject to a range of interpretation, from a requirement upon polluters to pay the costs of monitoring and management, through the requirement to pay the costs of clean-up, to the responsibility to pay for the cost of environmental damage as well as that of clean-up. Applying the principle may be simple or complex, depending on the nature of the environmental effects. Environmental economic assessment, and some form of economic planning instrument, are required in order to meet this principle.

Integration and Co-ordination

By definition integrated coastal management implies a greater level of integration than is typical of more conventional approaches. The rationale for this, especially in relation to aquaculture, has been clearly stated above.

Integration implies the involvement of a broad range of institutions in the decision making process. It implies the involvement of a wide range of stakeholders, with differing values, dealing with a wide range of development issues. It implies the facilitation of a broadened perspective on the part of these stakeholders. It implies an holistic analysis and synthesis of complex technical, social, economic and ecological information. It implies a correspondence between local initiatives and regional or national level policies, and vice versa. It implies better co-ordination between different sector policies. It implies increased cross-links between institutions, and/or new institutions.

Increased integration therefore implies increased complexity. Decision-making is likely to be slower and more difficult as the degree of integration increases. Key requirements are:

- high quality, well presented and effectively communicated/exchanged information;
- clear and widely agreed decision criteria;
- clear and transparent decision making processes; and (if necessary)
- a clearly designated (and widely agreed) final authority and arbiter (whether individual or committee).

In the case of less ambitious local or enhanced sectoral initiatives, these problems are reduced, but the potential for real integration more limited.

Public Involvement

Significant public involvement is a desirable and necessary part of any planning initiative. It takes different forms, all of which are important:

- communication of information: from decision makers, planners or technical specialists to other stakeholders, and vice-versa;
- participation: shared responsibility and decision making

Comprehensive public involvement is increasingly emphasised in any assessment or planning exercise for the following reasons:

- planning makes assessments and judgements about issues of widespread public concern: the quality of life, the value of resources, and the trade-offs between different resource uses;
- many of the assessments are subjective, and can only be agreed and/or validated through the widest possible consultation;
- local people can provide essential information about local natural resources, their status, use and value (sometimes referred to as indigenous technical knowledge);
- early exchange of views on key issues allows for the identification of information needs and improved focus of survey or research;
- public involvement may also reduce conflict through the early identification and resolution of potentially contentious issues;
- widespread consultation may generate new ideas for development alternatives, and possibilities for zoning/siting, design, and mitigation of adverse environmental effects;

- the more participatory forms of public involvement allow otherwise under-represented groups access to the decision making process;
- public involvement may contribute to, and/or reduce, the costs of monitoring and quality control;
- by engaging all stakeholders in the evaluation and decision making processes, it creates a sense of accountability, ownership and responsibility; and
- it increases transparency and accountability in decision-making, and ultimately increases confidence in decision makers.

Assessment of Costs and Benefits

The costs and benefits associated with aquaculture or other developments are rarely assessed objectively or comprehensively. If planning is to have any success in terms of optimal resource allocation to aquaculture and other activities, assessments of this kind must be thorough.This principle implies the need for risk assessment, as well as comparative and environmental economic appraisal of the full range of alternative development options or strategies, including non-aquaculture developments.

Estimation of Environmental Capacity

Environmental capacity is much quoted in relation to coastal aquaculture development. Although it is often difficult to quantify accurately, the concept is useful, and serves as a framework for the discussion of issues such as environmental standards, ecological processes, and the environmental values and perceptions of different stakeholders.

Emphasis on Incentives rather than Constraints

Regulatory approaches to the planning and management of aquaculture development often have limited impact, especially where aquaculture is small scale and widely distributed. Policing is in many cases difficult, costly, and unpopular. It may be made more effective if responsibility for design, implementation and enforcement is located at the proper administrative level, and full use is made of self-management and self-enforcement capacity by industry and farmers' associations. Incentives, on the other hand, do not suffer from the problems of evasion and non-compliance, and in some cases can be used to stimulate innovation leading to more environmentally friendly technologies. The use of economic instruments to influence both siting and operation holds considerable promise. Although some positive incentives may be costly, it should be possible to pay for them with negative incentives (e.g. taxes on undesirable locations, activities, technologies). However, regulation may nonetheless be necessary, and a balanced approach is required.

Control of Effects rather than Scale of Activity

Many forms of regulation of aquaculture (and indeed other coastal activities) are related to scale—either the area of land or water directly used by aquaculture, or the total production. For example, in line with the precautionary principle, an upper limit may be placed on aquaculture production in a bay, estuary or lagoon.

This limits the potential economic development, while providing no incentive to improve the environmental efficiency of the operation. It would not, for example, provide an incentive for the use of low pollution diets. A limit on effects (for example the concentration of nitrogen in the water at critical times of year) would provide an incentive for improved environmental efficiency through technology or management, while also allowing for increased production. However, there are some difficulties with this approach. The relationship between cause and effect may be only partly known or understood especially where multiple uses of the resource already exist.

Effective Institutions and Representative Organisations

The importance of institutional structures, roles and capacity cannot be over-emphasised. However, it is impossible to provide simple prescriptions for appropriate institutions and procedures without reference to specific contexts. An important part of any integrated planning initiative should therefore be an institutional analysis.

Institutional change is difficult and risky, especially where many different organisations are involved. Significant structural change should therefore be approached with caution. More modest changes to procedures, both within and between existing institutions, may be the most appropriate first step. Once the initiative is underway, the need or otherwise for more institutional change can be assessed.

Effective stakeholder representative organisations should facilitate the process of public involvement. Furthermore, such organisations may take active roles in the exchange and dissemination of information, and policy development. They are particularly important for minority or highly dispersed groups whose opinions may be difficult assess in public meetings or through survey.

LEGAL AND INSTITUTIONAL FRAMEWORKS

The importance of legal, procedural and planning frameworks designed to facilitate sustainable aquaculture development is emphasised in the FAO Code of Conduct for Responsible Fisheries:

> States should establish, maintain and develop an appropriate legal and administrative framework, which facilitates the development of responsible aquaculture.

> States should produce and regularly update aquaculture development strategies and plans, as required, to ensure that aquaculture development is ecologically sustainable and to allow the rational use of resources shared by aquaculture and other activities.

The need for a clear and comprehensive legal framework has been explicitly recognised by all those countries that have become significant producers of farmed shrimp. At the FAO Technical Consultation on Policies for Sustainable Shrimp Culture the following recommendation was made:

> "Governments should have a legal framework which applies specifically to coastal aquaculture, including shrimp culture"

and appropriate objectives for such a framework should be to:

- "facilitate and promote the development of sustainable aquaculture practices;
- promote the protection of coastal resources;
- promote the contribution of aquaculture to food security, national and international wise."

The approach adopted will depend on existing laws, traditions, and institutional structures. For example, an enhanced legal and institutional framework to promote planning for sustainable coastal aquaculture development could be built up around existing legislation and/or procedures for:

- sector planning;
- regional or district planning;
- watershed or coastal zone planning and management;
- environmental assessment.

Where the introduction of new legislation is difficult, or will cause excessive delay, guidelines for developing new initiatives may be introduced prior to specific legislation, as a means of testing out different approaches.

PLANNING PROCESS

Main Stages

The planning process is broadly similar irrespective of the administrative level at which it takes place, or the degree to which it is integrated. A major international workshop on ICM in tropical countries, held in 1996 in Xiamen, China P.R., discussed lessons learned from successes and failures experienced with ICM efforts. The workshop generated an overview of the processes of formulating, designing, implementing and extending ICM within the East Asian region as well as to other regions, and a set of Good ICM Practices. This agrees closely with the outline presented by GESAMP. It is applicable to local and enhanced sectoral initiatives as well as to more integrated approaches. The whole process should be seen as part of a dynamic and repetitive cycle leading steadily toward more sustainable forms of coastal development in general, and coastal aquaculture development in particular.

Operational Components

The first stage (stage setting and planning) can be broken down into a more detailed set of operational components, each of which may draw on a range of tools. These components are not necessarily in chronological order. Indeed, it is highly desirable that components 1 to 5 take place in parallel, since each should serve to inform the scope and focus of the others. The tools associated with each stage may or may not be used according to local circumstances and the scope of the initiative. It is apparent that the inclusion of all these components in any planning initiative is a formidable task. Attempts at more integrated planning for aquaculture development, and indeed ICM initiatives in general, have rarely encompassed all these elements. However, if the ideals of sustainable development are to be realised, the main operational components must be included, and facilitated where appropriate by the various tools available. The process can be made more manageable, and the research/information collection more focussed, if the principles of public involvement, evaluation, iteration and adaptation are applied at all stages of the process. Massive technical research and assessment exercises, however thorough, are unlikely to provide a sufficient basis for solving resource management problems, and should be used rather as an on-going input to a rolling and adaptive planning process.

Identifying the Mechanism and Level of Planning

Any attempt to improve the planning and management of aquaculture, either from a sectoral perspective or as a component in a broader integrated management exercise, must be initiated with great care. Such initiatives are likely to be seen as threatening to some stakeholder interests. The "who" and "how" of initiation can have a major impact on long term success. There are four key elements or pre-conditions:

- a policy or legal framework which requires, facilitates, or (at minimum) allows for improved and more integrated planning and management;
- an organisation or body which can lead or oversee the planning process, which is widely respected and trusted by the stakeholders, and which can effectively "deliver" on the guiding principles.

- an awareness of the need for improved planning and management on the part of the stakeholders; and
- adequate funding and/or staff time to undertake the process.

Where these conditions do not already exist, they must be created before any form of improved planning or management can be initiated.

A review of relevant policy and legal framework should be undertaken to determine what is possible, and how the initiative is to be taken forward. In the case of enhanced sectoral initiatives the analysis would normally identify the most appropriate mechanisms for consultation, exchange of information, and policy integration within the existing legal/institutional framework. It would also consider how the outcomes of the initiative could be implemented.

In the case of more ambitious and integrated initiatives, it may be useful to undertake institutional analysis and a preliminary stakeholder analysis to help in the choice or definition of an appropriate body to co-ordinate and oversee the planning and management process. This may be some form of committee, working group, or steering group including representatives of relevant government departments and agencies, industry and user groups and community representatives. It must be able to integrate technical and socio-economic information, including the needs and aspirations of stakeholders, and co-ordinate or implement interventions to meet coastal and aquaculture management objectives. This body would normally be convened and chaired by local or regional government, depending on the scale of the planning exercise. The main tasks and responsibilities of the co-ordinating body would be to:

- clearly delineate the responsibilities of all involved in the different stages of the planning process, especially with regard to implementation;
- facilitate, and set in place procedures, for consultation between agencies and between agencies and other stakeholders;
- set in place procedures for the exchange of relevant information between different interests;
- set in place procedures for defining and implementing planning interventions;
- set in place procedures for monitoring the effectiveness of interventions in terms of overall ICM objectives;
- set in place procedures for adapting interventions in the light of experience;

Clearly funds must be identified, preferably from a range of cross-sectoral sources to maximise ownership of the initiative. In some cases a re-allocation of staff time may be all that is required for the initial stages. Identification and raising of further funds would also normally be a responsibility of the co-ordinating body. In principle however, once the planning and management system is set up, and in the long term, it should be largely self-financing.

Commitment of Key Stakeholders

The trust and support of key stakeholders is essential from the outset. This can be gained through consultation and public involvement and is closely related to the establishment of the co-ordinating or advisory body, and the openness of its deliberations. Trust and support is likely to be lacking if the need for improved planning and management is not widely accepted, particularly by aquaculturists themselves. A series of public meetings and/or focus group discussions related to problems and potentials in aquaculture development, with a strong emphasis on more strategic development, may help gain support for the process, and serve as a useful input to any stakeholder and institutional analysis.

Understanding the Development Context

Key issues for improved management of coastal aquaculture may be social, environmental, technical, or economic. Identifying these issues implies a thorough understanding of both the development context (natural resources and ecology; human resources and economy), and the nature of actual and potential activities or developments (technical, economic, social and environmental characteristics). This can only be done effectively using an iterative and adaptive approach:

- assimilate existing information;
- identify key issues;
- identify further information and research needs;
- collect information and undertake research;

  refine key issues; etc.

Public involvement

Four basic approaches/tools can be used to identify key players, collect information, identify issues and possible conflicts, and encourage participation and ownership:

- social survey, supplemented with public information campaigns and limited public meetings;
- rapid appraisal (relatively informal but structured interviews and discussions with a wide range of stakeholders to gain information and understanding);

- participatory appraisal (wide-ranging exchange of views and information, with direct involvement of stakeholders in the decision making process);
- stakeholder consultative committees

The first approach is the most traditional, but is difficult, costly, and sometimes misleading. The second and third approaches avoid many of the problems associated with social survey, especially in relation to values and quality of life issues. Rapid appraisal is by definition rapid and relatively cheap. Participatory approaches, since they directly involve people in the planning process, should lead to a greater and wider sense of ownership and responsibility, and therefore increase compliance with planning provisions or regulations at a later stage. However, participatory approaches have their own problems, including the cost and time required, and the difficulty of involving all relevant stakeholders.

More effective participation of coastal resource users may be facilitated through the establishment of user groups or organisations to represent particular interests on consultative committees, or in higher level decision making processes. It is notable that an ICM initiative in Ecuador, by setting up consultative committees of resource user representatives, actually stimulated the establishment of user organisations. It is also notable that in Thailand the establishment of shrimp farmer associations has greatly enhanced shrimp farmer representation at meetings and on committees charged with developing aquaculture development policy. Even where a significant new planning and management initiative is lacking, the establishment of consultative committees may provide a basis for enhanced planning and decision making through existing procedures.

Description and mapping

The collection and assimilation of data relating to natural and human resources has received widespread attention, and has been facilitated through the use of remote sensing (RS) and geographic information systems (GIS). The latter can be used to assimilate effectively pre-existing information (for example relating to soil or water), any new information collected using the social survey techniques described above, or any more specific information collected in response to identified research needs. GIS can also be used as an accessible database for monitoring information both before and during plan implementation. The scope of GIS is usually restricted to physical parameters, but attempts have been made to extend it to financial and economic parameters.

In the case of aquaculture, this stage or component is commonly closely linked to issues of site suitability and possible zoning.

It is essential that mapping, description, RS and GIS are carefully focussed, and undertaken in parallel with, and guided by, discussions relating to issues identification and the setting of goals and objectives. This is an example of the iterative and adaptive process described above: RS and GIS may help in the identification and clarification of key issues; but equally, the identification and analysis of key issues (e.g. through public consultation) should serve to define the focus and scope of RS and GIS. This can then further illuminate the key issues. Without constant feedback and adaptation of this kind, RS and GIS can get out of hand, and become themselves costly ends, rather than planning tools. There must also be a clear rationale and mechanism for their use and maintenance.

Understanding physical and ecological processes

The coastal environment is dynamic. A static description of resources must be supplemented with an understanding of processes, dynamics, and interactions. This is particularly important for coastal aquaculture, which is often dependent on tidal regimes and hydrography, plankton communities and water quality, and soil, water and nutrient fluxes.

In many cases a descriptive synthesis of existing knowledge will be adequate in the first instance. In the longer term, and relating to critical components in the system, it may be useful to develop physical models (such as nutrient flux models), ecological models, and ultimately systems models.

The development of the more sophisticated of these models is usually difficult and time consuming, and cannot be considered as a pre-condition for either enhanced sectoral or fully integrated coastal management. Rather, the scientific research, and any associated modelling effort, must evolve and focus in parallel with, and informed by, participatory approaches to issues and problem identification. Once a plan is implemented, thorough and regular evaluation should help identify or focus specific research and modelling needs, which can then be used as a direct input to the planning and management process.

Where physical or ecological issues are complex, or modelling capacity limited, it may be more convenient and cost effective to invest in an environmental monitoring programme, with basic parameters to be recorded (related to environmental objectives discussed below), and clearly defined (and widely agreed) response procedures.

Environmental capacity

Environmental capacity is a key concept in the idea of sustainable development and must therefore be addressed in any initiative designed to promote sustainable development. Environmental capacity (otherwise referred to as assimilative capacity) is "a property of the environment and its ability to accommodate a particular activity or rate of an activity...without unacceptable impact". Different mechanisms of impact can be identified for a particular situation, and the capacity of the environment to absorb each of these can be estimated.

Environmental capacity measures the resilience of the natural environment in the face of impact from human activities, and must be measured against some established standard of environmental quality. Understanding and measuring environmental capacity allows for the determination of the scale of activity (using a specified technology) which can be accommodated without threat to an environmental standard.

In the case of aquaculture, environmental capacity in relation to a specified area might be interpreted as:

- the rate at which nutrients can be added without triggering eutrophication;
- the rate of organic flux to the benthos without major disruption to natural benthic processes; or
- the rate of dissolved oxygen depletion that can be accommodated without causing mortality of the indigenous biota.

Understanding Options

A comprehensive description of the technical, economic and resource use characteristics of different technical production systems or species is a prerequisite for the rational appraisal and comparison of the sustainability of different development options. It is essential for the clarification of key issues, and as a basis for any kind of strategic planning or market intervention. The objectives for such an assessment are:

- to identify financially viable aquaculture and alternative production systems (in the short, medium, and long term);
- to provide resource use/transformation information which can be used for accurate environmental and social assessment, and cost benefit analysis;
- to provide "sustainability" profiles of alternative development options (summaries of social, economic and environmental characteristics) which can be used to inform participatory discussions and decision making.

Ideally such assessments would include, for each type of development

- screening for technically feasible development options;
- market assessment;
- a description of suitable location and site requirements;
- standard financial analysis, investment appraisal and sensitivity analysis;

- a profile of resource use and waste output, including temporal or seasonal variations;
- a profile of socio-economic characteristics, including potential employment generation, labour seasonality, income generation and distribution, and barriers to entry (skills, capital, natural resources requirements);
- risk analysis
- synthesis—comparative economics and technical appraisal of different options;

This kind of information is commonly generated in respect of individual projects through feasibility studies and investment appraisal, although resource use and socio-economic characteristics are rarely specifically or adequately addressed. Nor are these studies commonly applied to a whole sector, or to a range of possible development options. Information is therefore usually inadequate for comprehensive and *comparative* environmental assessment, cost benefit analysis, and other formal decision making processes or techniques.

Each development option should also be assessed in terms of how, and to what extent, it might contribute to the objectives of the planning and management initiative. This will be easier if clear criteria for measuring progress against objectives have been agreed.

Identifying Development Priorities

The information collected and assimilated, relating to both the development context and the nature of existing activities or development options, should provide a sound basis for the assessment of development alternatives (activities, technologies, operational practices) against planning objectives, standards and targets, and the selection of preferred options. This assessment should identify:

- undesirable or low priority activities/developments;
- desirable or high priority activities/developments;
- design, technology and management practices likely to maximise social, economic and environmental benefits from different activities;
- mitigation measures to minimise any adverse social or environmental impacts related to otherwise desirable activities or developments;

There are three main techniques or approaches for presenting, analysing, comparing and prioritising the alternatives:

- Environmental assessment;
- Cost Benefit Analysis;

– Participatory decision making

Ideally, all three approaches should be used, each informing the other.

Environmental Assessment (EA)

Environmental assessment is a comprehensive format for the assessment of a development or set of developments. It brings together information about the development context and development issues with the technical-economic appraisal, allowing for a comprehensive assessment of the environmental and social effects of a proposed development, or development option, in a particular location. It has become a standard planning tool, with a significant impact on development decisions, and has been used widely in coastal management. In recent years increasing emphasis has been placed on integrating social and economic considerations into the EA process.

Unfortunately EA has commonly been applied on a project basis, and normally only to larger projects. Although such an approach is useful, and indeed essential for large business or infrastructure, it cannot address the incremental but cumulative social and environmental problems associated with large numbers of small agricultural or aquaculture developments. It has also often been applied in the absence of clear and agreed criteria against which impacts can be assessed.

To be effective as a tool for more integrated coastal management, EIA should be applied to all sectors in the form of Integrated Environmental Assessment· If this is beyond the scope of the planning exercise EA should be applied to the whole aquaculture sector, and should be used as a tool for strategic planning rather than as a regulatory instrument. If the resource and technology assessment as described above has been done thoroughly, this process should be relatively straightforward. The likely environmental and social impacts of a range of technologies or development options in different locations can be compared, and planning interventions to minimise environmental impact devised. Sector environmental assessment has the potential to become a standard tool for strategic planning.

EA typically assigns "significance" to social and environmental impacts. The level of significance is typically used as a decision criterion. It is therefore important that significance itself is measured against some consistent and objective standard. This implies the existence of social and environmental standards or targets against which significance can be measured. The principles and practice of EA are similar whether applied to individual projects or to a "sector" and are now well established.

Economic analysis

Economic analysis provides a formal framework for comparing the multiple functions and uses of the coastal zone. In particular, it allows for a systematic evaluation of alternative uses, which in turn can lead to the identification of options for planners and decision-makers with regard to the allocation of resources. It may also provide a framework for a rigorous assessment of the costs and benefits of interactions between activities. In its more sophisticated forms it seeks to aggregate social, environmental and financial values relating to a particular enterprise or sector into a single index of total economic value, allowing for a standardised comparison of development alternatives. Cost Benefit Analysis provides an overall framework for the comparison of development options and is widely described in the literature. Environmental economic techniques allow for the inclusion of non-market or traded goods and services.

Although there are many advantages in bringing together all the diverse values within a single overall framework, these approaches should be treated with caution. They commonly require complex and sometimes questionable sets of assumptions. They should only be used when the methodology and underlying assumptions can be clearly explained to decision-makers. This may be difficult in the case of more participatory decision making, involving many individuals and interests with radically different educational levels.

Where comprehensive CBA and environmental economics are not used, simpler economic analyses can nonetheless shed light on specific issues such as the "costs" of pollution or habitat degradation associated with particular activities. Where such figures are difficult to estimate, it may be possible to provide information on the "opportunity costs" associated with *not* developing or polluting resources. In other words simple economic analysis can provide information on the nature of the "trade-off" between competing uses of coastal resources, which is vital for economically sound and well informed decision making.

Participatory approaches

There are a variety of decision-making techniques that avoid the problems of aggregating dissimilar values and quantities noted above. These rely on meetings, discussions or interviews to determine values, and examine the trade-offs between different options. For example, some options may yield higher environmental benefits, while others yield higher financial benefits. Decision makers will generally seek to balance the two, essentially trading environmental benefit for financial benefit up to a point which reflects their relative valuation of financial and environmental benefit.

There are a variety of group/participatory tools that facilitate decision making of this kind, commonly referred to as multi-objective decision analysis Most are based on ranking (which is usually much easier than assigning absolute values), or paired comparisons of development options against different decision criteria. Financial or money values are used to a variable degree according to the nature of the alternatives. These approaches rely heavily on:

- effective communication of the various characteristics or possible values and impacts (economic, social, environmental) of different development alternatives—and the trade-off between them (i.e. how much is lost or gained of different benefits as the options are changed);
- clarification of development/environmental objectives and targets, so that development alternatives can be measured (and therefore valued) against them.

The use of such approaches necessarily encourages a more integrated assessment, and allows for a wide range of stakeholders to make an input to the decision making process.

Transparency

Given the technical difficulty and cultural subjectivity of assigning values, especially in respect of social and environmental issues, the key to both approaches is *transparency*. The assumptions lying behind any aggregate figures or trade-off calculation must be clearly understood by those involved in the decision-making process.

Conflict Identification and Resolution

Conflict between different resource users may already exist in the coastal zone. Indeed, an integrated planning initiative may be a response to this conflict. Where conflict does not already exist, it may actually arise during the planning process. Public involvement and participatory decision making, environmental assessment, cost benefit analysis, and other techniques which seek to identify and compare social, economic and environmental values, may bring into the open previously hidden differences in terms of development needs, and values and aspirations between different resource users and other stakeholders. For example, public consultation in relation to shellfish farming in Sweden, and shrimp farming in India and East Africa has led to serious conflict.

Clearly this kind of conflict should be avoided as far as possible, and the way in which issues are presented for public consultation should be handled with great sensitivity. One important rule is to establish broad public agreement on overall development objectives, strategy, and decision criteria, before addressing specific development cases or projects. Trying to establish decision criteria on an ad hoc basis in relation to individual projects will inevitably polarise opinion.

This reinforces the need for comprehensive stakeholder participation in the initial setting of planning objectives, which should reduce the likelihood of conflict arising during the formulation and implementation of the plan. It may be that the outcome of any conflict resolution is in fact a tightening or reformulation of the planning objectives.

Should conflict nonetheless arise, a variety of approaches are available for its resolution , including litigation, arbitration, mediation, and negotiation. The second two are usually more desirable than the first two. Litigation and arbitration both involve the imposition of a solution, and may not address or resolve the underlying causes of conflict, which may therefore re-surface at a later date. Mediation and negotiation on the other hand seek to resolve differences through an emphasis on common objectives.

Defining Broad Management Strategy

The process of public involvement, agreeing broad objectives, and evaluating alternative development (and conservation) options, should provide a comprehensive information base for defining an overall planning strategy or framework.

Sector/activity targets

A significant part of strategic planning is the setting of targets relating to particular sectors or activities. Sector targets may be set in terms of *total output* (production and/ or value), or *total allocation of resources* (e.g. zoning of land or water; allocation of environmental capacity). They may be set in relation to an entire coastal area, or in relation to specific zones as defined above.

Output targets have commonly been associated with planning in centrally planned economies, but with rather little emphasis on land or resource use, environmental effects and environmental capacity. In contrast, they have generally had a minor role in integrated coastal management, where the emphasis is typically on the resource base and environmental conservation.

Environmental capacity assessment attempts to define the relationship between sector activity levels and environmental quality, and should therefore be of particular interest in strategic conservation planning. It is an important tool for integrating aquaculture into broader coastal planning and management initiatives. Output targets may be associated with a resource allocation target—for example a particular level of production from a specific zone.

Criteria for locating aquaculture

A key component in the integration of aquaculture in strategic planning is an understanding of the specific needs of aquaculture in terms of site and locatioin requirements. These should have been defined during the technical and economic assessment phase of the planning process. These requirements may be developed into a set of criteria which can be used as the basis for a variety of planning and management interventions (such as zoning, or screening development proposals) to influence location and siting of aquaculture activities. Appropriate siting of aquaculture and other development activities in coastal areas

- minimise the risks to, and maximise the returns from, coastal aquaculture;
- maximise the overall economic return from all activities in the coastal zone;
- minimise conflict between aquaculture and other resource uses; and
- minimise environmental impact.

Criteria for assessing aquaculture

A variety of other criteria for assessing the social, economic and environmental impact of individual projects or proposals should be developed, based on the planning objectives. They might, for example, relate to scale, design, technology, management, or labour relations. These might be used in support of the various planning instruments described below. For example, the assessment criteria could be used as the basis for credit or grant aid, for the issuing of permits, for screening proposals for environmental assessment, or as standard criteria to be used when undertaking environmental assessment.

Zoning

Zoning implies bringing together the criteria for locating aquaculture and other activities in order to define broad zones suitable for different activities or mixes of activities. GIS is particularly well suited to facilitating this task.

Zoning may be used either as a source of information for potential developers (for example by identifying those areas most suited to a particular activity); or as a planning and regulating tool, in which different zones are identified and characterised as meeting certain objectives. Zoning of land for certain types of aquaculture development may:

- help to control environmental deterioration at the farm level;
- reduce adverse social and environmental interactions;
- serve as a focus for estimates of environmental capacity.
- serve as a framework for the provision or improvement of water supply/drainage infrastructure to small scale farmers.

The strength of zoning lies in its simplicity, its clarity, and its potential in terms of streamlining procedures. For example, once a zone is established and objectives defined, then developments that meet the objectives and general conditions for the zone may need no further assessment. What is allowed and what is not allowed is clear, and developers can plan accordingly. Any monitoring required can be applied to the whole zone rather than individual farms.

Its weakness lies in its rigidity. No zone is perfect, land/water capability assessment may have been inadequate, boundaries are frequently arbitrary, and conditions may change. There may be small pockets of land or water of high potential for aquaculture, which were not recognised in the resource assessment process, which are not part of an aquaculture zone, and which are therefore prevented or subject to severe regulation. In some situations this could restrict access of poor people to the opportunities for aquaculture development. Furthermore, it may actually be undesirable to encourage a concentration of aquaculture in one particular area, however suitable it may be, because of the increased risk of rapid spread of disease.

Planning Instruments: Incentives and Constraints

A wide range of actions may be undertaken to influence the nature, direction and location of development, so that the strategy can be implemented, and the planning objectives met. To be effective, these must be capable of serving as incentives for, or constraints to particular activities, forms of development, or the location of development. They might include:

- infrastructure development;
- training, education and awareness raising;
- economic instruments;
- regulatory instruments;
- codes of practice;
- markets and labelling; and
- improved institutional linkages

Many of these instruments work best in combination. Infrastructure, tax incentives, and regulations may all work together to achieve some particular objective. They may be associated with zoning to provide differential incentives (e.g. applied in some zones but not in others) and thus reinforce the zoning system.

Infrastructure and services

Infrastructure development may be used to improve the performance of aquaculture in particular locations, minimise environmental impact, and reinforce zoning by offering a comparative advantage to farmers who operate within the zone For example, new or improved water supply and wastewater treatment for aquaculture may make a significant contribution to sustainability. They may:

- reduce problems associated with existing aquaculture developments;
- pre-empt such problems arising in relation to new developments; and
- influence the location of new aquaculture development, with a view to ensuring that it is located in areas where it is most likely to succeed, and where conflicts with other activities or resource users are minimised.

The provision of improved marketing and processing facilities, or the provision of infrastructure to stimulate more rapid development of private sector services, may also have a significant influence on the success and location of new aquaculture development. Access to credit may also be considered as a form of infrastructure or service.

Training, education and awareness raising

There is growing awareness of the importance of environmental issues, but this awareness is "patchy" and may be inadequate in areas experiencing very rapid development. Raising environmental awareness will reinforce the impact of specific incentives or constraints. Research, extension services and training can be used to influence the sustainability of the sector, by offering information and advice about environment friendly technology and practice, or advising on site selection.

Economic approaches and instruments

There is increasing interest in the use of economic incentives and constraints to promote environmentally and socially friendly aquaculture siting, design and operation. Most measures of this kind require legislation and implementation at provincial or national government levels. Economic instruments include:

- access charges;
- pollution charges;
- tradable permits for resource use, harvesting rights, pollution/emission rights;
- various forms of subsidy and/or credit for environment friendly location, technology or management;
- refundable deposits and bonds laid against possible environmental damage, or set aside for restoration purposes;
- process or product standards linked to labelling and marketing initiatives.

Economic instruments appear to have many strengths. They:

- can be used to directly implement the polluter/user pays principle;
- require little in the way of enforcement;

- promote innovation;
- are flexible and efficient;
- may not require farm specific information on operation or discharge;
- can be linked to environmental capacity;
- can be used to address cumulative problems; and
- can generate government revenue for environmental management.

They also have weaknesses:

- the actual effects on farmer behaviour are not very predictable;
- they may need sophisticated institutions (to define; monitor; adjust, adapt etc);
- they are not always popular with government agencies, since they imply less direct control;
- they are not always popular with industry, since (negative instruments) imply extra costs.

To date these instruments have been little used to influence the course of aquaculture development and deserve serious consideration. It is probable however that regulatory instruments will still be required to complement these approaches.

Regulatory instruments

Regulation has been commonly used in an attempt to manage the development of and impacts from aquaculture. This has succeeded in many instances, especially in developed countries, but has a rather poor record in many developing countries where the enforcement of regulation may be particularly difficult. Regulation may include any or all of the following:

- restrictions on location. For example some zoning schemes may explicitly bar certain types of development or activity;
- prohibition of specific activities, materials or technologies;
- requirements for specific activities, technology or design;
- effluent standards;
- receiving water standards;

These may be stand-alone regulations, or may be directly linked to registration, licensing or the issue of permits. Many regulations are difficult to implement in practice, and may lead to an attitude of limited responsibility by the farmer. Their use should be limited as far as possible, but they remain an important last resort, and may serve to reinforce more positive incentives and economic instruments.

Codes of practice

Where the rationale for regulation is clear, and particularly when it relates to the interests of farmers themselves, every effort should be made to promote self-regulation through codes of practice. These may be reinforced through peer pressure, and in some cases actually enforced by associations of farmers themselves. In Thailand for example, the Surat Thani Shrimp Farmers Association has agreed to its own set of standard procedures related to water quality and disease management for a group of neighbouring farms with common water supply. Codes of practice, including best management practice, may be used as a basis for certification and quality labelling.

Markets and labelling

Farm gate price has a major impact on farmer behaviour. If this price can be linked in any consistent way to better siting and management of aquaculture operations, change will follow rapidly. Quality or "green" certification and product labelling offer significant opportunities, albeit relatively untested, in this regard. However, the benefits from such schemes may accrue largely to the wholesaler or retailer rather than the producer, and directly linking farm gate price to environmental management will not be easy. Furthermore the process of certification is difficult and costly, and the impact on price will depend on consumer trust of the certification system. These approaches may also be easier for larger scale and well organised producers, with consequent difficulties for small scale producers.

INSTITUTIONAL AND STAKEHOLDER ANALYSIS

Institutional Analysis

Institutional analysis is an essential part of any new planning and management initiative, especially where a greater degree of integration is sought. The nature and operation of institutions, and their mode of decision making, will have major implications for the implementation of any strategy or planning related to the promotion of sustainable development. The nature of the existing institutions should therefore be assessed, and new institutions, or frameworks for institutional collaboration and joint decision making, should be established if necessary. Great care should be taken with "model" institutional frameworks for more integrated planning. An institutional structure transformed to a new cultural context is unlikely to operate in the same way.

Institutional analysis covers both formal and informal institutions. Formal institutions are those such as government agencies, and they typically have a legally defined role, structure, and in some cases, sets of procedures. Informal institutions are those such as business, social or family networks or associations. The latter also have structure and sets of procedures, although these may have no legal or written basis. In either case institutional analysis requires that both structure and procedures are described and analysed. In essence this requires that the following questions be addressed in relation to any planning issue (such as land or water use):

- What are the rules?
- Who decides, and how (i.e. process and decision criteria)?
- Who implements, and how?
- How and when is progress assessed?
- What are the relationships between different institutions (formal and informal)?

The main types of institution which are likely to be relevant to planning and management of aquaculture are:

- Local, district, provincial and national government (formal);
- Agencies and advisors of government;
- Formal and informal business associations;
- Non-governmental organisations (NGOs);
- Religious institutions;
- Town, village or commune decision making structures (formal and informal).

Townsley presents a summary of the institutions and levels to be considered and the specific tools which can be used to analyse them. Pido *et al.* present a set of guide questions which can be used as a framework for research, and as a basis for discussions and interviews.

Stakeholder Analysis

Stakeholder Analysis is related to institutional analysis, but places far more emphasis on individual motivation and/or collective interest, than on structures and procedures. It may be summarised as an approach to understanding a system, and changes in it, by identifying key actors or stakeholders, and assessing their respective interests in that system. It is of particular importance where competition for, or depletion of natural resources is an issue, and is therefore of great relevance to planning for sustainable aquaculture development. It is a necessary starting point for any kind of public involvement, since it will help to define who needs to be involved, and how.

Stakeholders are all those who have interest in the issues being addressed—as measured in terms of welfare or utility. Some are active—they affect the system; some are passive—they are affected by it. They may be considered in detail because of their:

- importance—as possible beneficiaries of development;
- influence—their power over the success of project;
- or because they can be identified as winners or losers in the development process.

Stakeholder analysis seeks to:

- identify, assess and compare their sets of interest;
- examine inherent conflicts and/or compatibilities, and
- describe and explore trade-offs.

Exploring trade-offs (e.g. short term benefits of income from habitat destruction and conversion, versus long term benefits of biodiversity, storm protection etc.) as perceived by different stakeholders, is of particular importance. It can be used as one method for assigning values and/or ranking objectives and development or management alternatives.

PUBLIC INVOLVEMENT

The widest possible range of stakeholders should be consulted or actively involved in the planning process, ideally following on from a stakeholder analysis. At minimum, the following should be actively involved:

- existing and potential fish farmers and other aquaculture industry representatives;
- the administering (sectoral) agency;
- the people, individuals or groups in local (affected) communities;
- specialist government agencies;
- NGOs and technical specialists;
- others, such as donors, banks, development agencies, academics

Whatever the particular mix of techniques used which may be suitable according to local circumstances, it is important that some basic principles are adhered to:

- sufficient relevant information should be provided in a form that is easily understood by non-experts early in the planning process;

- sufficient time should be allowed for stakeholders to read, discuss and consider the information and its implications;
- sufficient time should be allowed to enable stakeholders to present their views;
- the selection of venues and the timing of events should encourage maximum attendance and a free exchange of views by all stakeholders (including those that may feel less confident about expressing their views); and
- responses should be provided to issues/problems raised or comments made by stakeholders, so that confidence in the public involvement and planning process can be maintained.
- provisions should be made for updating the public and providing feedback on progress with the planning;
- more formal planning and management procedures should be integrated with more traditional decision making processes where possible

A variety of relatively standard techniques are available to collect information, identify issues and possible conflicts, and encourage participation and "ownership".

Socio-economic Survey

In some circumstances it may be appropriate to undertake a detailed socio-economic survey. There is a large literature on socio-economic survey techniques and questionnaire design, as well as training manuals, and these should be consulted prior to any major survey initiative. Full-scale social survey however should be approached with some caution. It is expensive and time consuming, and may be rather inefficient in terms of issues identification compared with the more flexible and participatory techniques described above. It is therefore best used to clarify or explore in more detail important issues that have been raised or identified using RRA or PRA.

Geographic Information Systems

Description and mapping is a basic starting point in the identification of many issues, especially with regard to resource use and allocation, and may also form the basis for specific planning interventions related to site selection criteria, and in some cases zoning.In recent years advances in Remote Sensing (RS) have greatly enhanced our ability to describe and understand natural resources and human activity. In parallel with these developments in information collection, the rapid development of Geographic Information Systems (GIS), has greatly enhanced our ability to store, analyse and communicate this information.

The scope of GIS is usually restricted to physical parameters, but attempts have been made to extend it to financial and economic parameters. There is little doubt that RS and GIS are useful tools, but they have sometimes been "technically driven" and taken a disproportionate share of the resources allocated to sector and integrated planning initiatives. They also have some limitations. Important factors which are commonly not taken into account in GIS and macro level site appraisal are the availability and cost of land (the major factor in site selection in practice), "micro" site features such as site water supply (canals, dykes, groundwater etc), and the diversity of soil conditions within broader soil type zones.

General Approach to Estimating Environmental Capacity

To be cost-effective, estimation of environmental capacity should only be applied to those forms of environmental impact likely to occur in a given situation. In principle, it need only be applied to the form that becomes limiting first. In practice this may be difficult to determine. A scoping exercise can identify relevant forms of impact with respect to the environment and technologies in question. In the case of intensive finfish and crustacean aquaculture for example, these will normally include the impacts of nitrogen, phosphorus, organic matter, and certain chemicals. In the case of shellfish (especially molluscs) they would include the reduction in the phytoplankton food source. Once this has been done, prediction of capacity follows in three phases:

- define acceptable limits of environmental change for a particular area or zone in terms of measurable variables ("measurement variables");
- define, and if possible quantify, the relationship between aquaculture (and ideally other activities) and measurement variables;
- calculate the maximum rate or level of activity which will not breach acceptable limits.

Establishing acceptable minimum or maximum limits for measurement variables (such as nitrogen concentration) should ideally be based on quantitative predictions of environmental consequences of the changes in these variables, such as destruction of organisms or habitats, eutrophication, or resource depletion to a level at which it becomes limiting to other users. Environmental standards related to these broader environmental features should have been agreed as a part of setting planning objectives and associated targets. Back-calculations can then be made to give acceptable levels for measurement variables. These are known as "effects based" standards.

In practice the relationship between measurement variables and environmental quality of relevance to the various stakeholders is often difficult to establish, and the measurement variables themselves are commonly used directly as the basis for environmental standards. These environmental quality criteria may already exist, derived for other purposes of environmental protection, but which must be adhered to for legal reasons.

If effects-based standards cannot be established, and existing standards are not available or relevant, it may be necessary to start with conservative values, which will provide a reasonable level of protection. These can be refined progressively once the procedure has been applied and its success monitored, as discussed below. Ideally they should be widely discussed and agreed with the stakeholders.

Clarifying the relationships between aquaculture activities (such as feeding), the measurement variables, and the environmental consequences, will depend on an understanding of physical, chemical and ecological processes including:

- the dispersal of nutrients (or other substances) in receiving water;
- the dilution of these substances in the receiving water;
- the degradation or breakdown of these substances in the water column or sediments;
- the adsorption of these substances by sediments;
- the assimilation of these materials by plants or animals;
- the effects of these materials on different components of the ecosystem

In practice the last four of these are complex and often ignored or approximated, while the first two are addressed using mass balance and dispersal models. These may be relatively simple or rather complex, depending on local hydrology.

REFERENCES

Asian Development Bank, 1991. "Environmental evaluation of coastal zone projects: methods and approaches". *ADB Environ.Pap.*, (8):72 pp.

Aguilar-Manjarrez, J. & Ross, L.G., 1995. "Geographical information systems (GIS); environmental models for aquaculture development in Sinaloa State, Mexico". *Aquacult.Int.*, 3(2):103-115

Chaston, I., 1989. *Marketing in fisheries and aquaculture.* Oxford, UK, Fishing News Books, 144 pp.

Dinwoody, C. & Teal, F., 1996. *Principles of cost benefit analysis for developing countries.* Cambridge, UK, Cambridge University Press, 302 pp.

FAO, 1998. *Report of the Bangkok FAO Technical Consultation on policies for sustainable shrimp culture.* Bangkok, Thailand, 8-11 December 1997.

11

Sustainable Agriculture and Rural Development

The concept of sustainable development is an evolving one, and there are many definitions in the literature, some very similar, and others markedly different. According to the World Commission on Environment and Development (WCED):

> Sustainable development is development that meets the needs of the present without compromising the ability of future generations to meet their own needs.

In 1988, and on the basis of the WCED's definition of sustainable development, the Food and Agriculture Organization (FAO) Council defined SARD as:

> ... the management and conservation of the natural resource base, and the orientation of technological and institutional change so as to ensure the attainment and continued satisfaction of human needs for present and future generations. Such sustainable development conserves land, water, plant and animal genetic resources, is environmentally non-degrading, technically appropriate, economically viable and socially acceptable.

Both of these definitions are strongly anthropocentric—the focus is on the sustainable welfare of humans. They contrast with definitions of sustainability proposed by some ecologists. As Schuh and Archibald note, it makes sense to concentrate on the welfare of people since any operational approach to the conservation of natural ecosystems must be rooted in the beliefs and values of society. Of course, those beliefs and

- sectoral polices at national level; and
- measures that can be taken to promote SARD at the local level.

Notwithstanding some wide acceptance of the WCED and FAO definitions of sustainable development and SARD, respectively, there is still plenty of disagreement about the definitions themselves and about their interpretation. Problems in finding an operational definition of sustainability are the main reason for some economists, such as Beckerman, arguing that sustainability is a meaningless notion. According to Pretty:

> ... any attempt precisely to define sustainability is flawed. It represents neither a fixed set of practices or technologies, nor a model to describe or impose on the world. The question of defining what we are trying to achieve is part of the problem, as each individual has different objectives.... Sustainable agriculture is... not so much about a specific farming strategy as it is a systems-oriented approach to understanding complex ecological, social and environmental interactions in rural areas.

In other words, Pretty was arguing that SARD is a learning process, not a goal—a view also held by some ecologists. Others have argued that sustainable development is about sustainable human livelihoods, particularly the relief of poverty now and in the future, implying a strong focus on current and inter-generation equity.

While the views of sustainability of people from different disciplinary backgrounds are often quite different, they are best seen as complementary, not conflicting. The three legs of the sustainability 'tripod' can be viewed as representing the economic, ecological and sociological schools of thought. Without all three legs the tripod will not stand. Each leg gives support to the others. Only if all three are firmly on the ground can the whole entity be strong enough to use.

SUSTAINABLE SYSTEMS

As Lynam and Herdt have pointed out, 'sustainability is first defined at the highest system level and then proceeds downwards; and as a corollary, the sustainability of a system is not necessarily dependent on the sustainability of all its sub-systems.' Sub-systems may be defined in terms of scope or scale. Thus, at the highest level, what matters is the sustainability of the planet earth. Invoking the Lynam and Herdt principle implies that individual sub-systems need not all be sustainable for global sustainability to be achieved, although the requirement of global sustainability constrains the tolerable levels of unsustainability of sub-systems. Within agriculture, clearly not all agricultural sub-systems may be sustainable.

In some cases, the prevailing system may be causing a progressive depletion of the natural resource base, for example, through soil erosion. In places where agriculture was traditionally based on shifting cultivation with bush fallow, population growth and perhaps the loss of access to some land may have led to an unsustainable reduction in the fallow length. Sometimes, in such cases, it may be possible for the system to evolve, perhaps with improved and appropriate technologies, into a sustainable form of settled agriculture. In other cases, where the natural environment is too fragile, there may be no technologies consistent with preventing the degradation of the resource base and providing all the people with sustainable livelihoods.

Such 'poverty traps' are all too common in various parts of the world. Sometimes it will be possible to find policy interventions to allow people to break out of such traps.

Examples include the introduction of improved technologies, perhaps combined with asset redistribution, or the development of associated rural industries.

On other occasions, no such answers may be found that could be implemented at reasonable cost, and it will be necessary for the excess population to move to other, less threatened areas. Resettlement schemes have been used to ease population pressure on land in a number of countries, such as Indonesia. While it is true that not all agricultural systems may be sustainable, now or prospectively, it is also the case, as indicated by Lynam and Herdt, that not all systems need to be sustainable.

Some substitution of resources may be desirable, even necessary, for overall sustainability to be attained. For a less developed country with relatively good forestry resources to impose on itself the condition that the harvest rate should be no higher than the regeneration rate may be a mistake. It would exclude the possibility of converting at least some forest areas to more productive agricultural use in crop or livestock farming. Some of the loudest calls for countries in the South to conserve their forests come from the North where large areas of forest and woodland were successfully converted to other uses long ago. It seems unreasonable to expect poorer less developed countries to deny themselves the same opportunities.

Depleting one natural resource may be consistent with the goal of sustainability in general, and SARD in particular, provided that the total stock of capital is not depleted. This might mean, say, investing in human capital to compensate for the unavoidable or considered depletion of capital inherent in a forest, a fishery or farmland. Much of the debate between economists and environmentalists about what constitutes sustainability centres on different perceptions of the extent to which such substitution is possible. Economists generally take an optimistic view of substitution possibilities and environmentalists a pessimistic one.

The need to take a global view of sustainability is also implied by the growing linkages between countries. With increase in international trade and greater mobility of people, capital and technology, it is no longer necessary for countries to aim for self-sufficiency. Indeed, to do so is to deny access to the advantages of international trade through exploiting comparative advantage.

TIME DIMENSION OF SARD

The FAO definition of SARD refers to the needs of present and future generations. The time horizon for planning and policy making implicit in such a definition is very long. However, because the long-term future is so hard to foresee, there are severe operational difficulties in setting a very long time horizon for policy making and planning.

Clearly, the choice of the appropriate time scale will depend on circumstances. Longer horizons may be unavoidable for long-term processes such as forestry while in other cases the here-and-now urgency of some environmental threat may need to be dealt with on a short-term basis. It is also important to distinguish between the time period for which plans are laid, herein called the planning horizon, and the time over which the consequences of those plans may be experienced.

Planning horizons of ten or fifteen years are usually about as long as it is plausible to consider. Even over such periods, the ability to account for changes in the technological, social, political and economic situation some several years into the future is very limited. In these circumstances, therefore, it is necessary to lay plans and set policies on a shorter term basis while giving due weight to the implications of today's actions for future generations. This requires the adoption of a sustainability principle in policy making and planning.

Adoption of a sustainability principle means making sure that actions taken now do not unduly limit the opportunities for people in the future to live as well or better than we live today. It means that we should not deplete the total resource stock that we pass on to the next generation. In fact, short of some massive calamity, it will take many years before global population growth can be halted and eventually reversed. Therefore we need to pass on the planet to the next more numerous generation in a more productive condition than it was when we inherited it.

The value judgements implicit in the ideas presented in the previous paragraph do not mean that we must pass on better stocks of all resources. To do so with the exhaustible resources such as fossil fuels is clearly impossible. Rather it is the total bundle of resources that needs to be improved from generation to generation. As one resource stock is run down, the total bundle can improve only if there are at least compensating improvements in the stocks of some other resources that can substitute for what is lost. In particular, human capital, embodied in such things as improved technologies and better management skills, can substitute, at least to some extent, for depletion of some natural resources. As noted above, there are differences of opinion about the extent to which such substitution will be possible in the future.

The concern for the welfare of future generations embedded in the SARD definition may appear to be in conflict with the practice of using a positive discount rate to bring future costs and benefits to present values when planning investments such as agricultural projects. Positive discount rates may seem to discriminate against future generations because projects with costs in the distant future but benefits in the near term will be favoured. On the other hand, investments that have long-deferred benefits will not appear attractive with discounting.

The negative impact on sustainability may be particularly strong for countries and individual resource managers who face high costs of capital. Heavily indebted nations may have limited ability to borrow internationally so that the actual or opportunity cost of capital is likely to be higher than less-indebted countries. Similarly, the productivity of capital for poor farmers in less developed countries may be very high because such people are often unable to access formal markets for credit. High interest rates may mean that long-term investments, such as land improvements, cannot be afforded.

Such reasoning has led some environmentalists to argue for a zero or negative discount rate to promote sustainability. However, it is not clear how actual interest rates could be forced down to zero without savage and politically unrealistic cuts in current levels of consumption. Nor is it clear that arbitrary use of a zero discount rate for public project appraisal makes sense. It would imply accepting projects showing low returns on investment when others would be available showing positive rates of return. To waste productive investment opportunities in this way would be to deny future generations the benefits from more productive uses of scarce capital. Moreover, lowering the discount rate may also lead to the adoption of investments that are resource degrading, such as logging low-yielding or difficult areas of natural forest, or fishing already over-exploited fisheries.

The consensus of opinion is that attempts to attain sustainable development by manipulating the interest rate are misplaced. It would be desirable to solve the problems of the excessive levels of indebtedness of some countries. Similarly, there is merit in promoting the development of financial systems to make it possible to deliver credit to rural resource managers who need it at reasonable cost. But tinkering with the discount rate to try to attain inter-generation equity is inappropriate—as is the not unheard of practice of government ministries deliberately selecting low discount rates in order to put pet projects in a better light.

A more satisfactory approach is to apply conventional investment criteria to all projects using normal discount rates, but also to invoke the sustainability principle in the analysis of public investment programs. That means selecting projects on the basis of discounted costs and benefits subject to the condition that the overall stock of capital (natural, artificial and human) across all investments is enhanced for future generations.

While the implementation of such a rule may be easier said than done, at least until much improved methods of measuring and monitoring changes in capital stocks are developed and brought into use, it is an important guiding principle for policy makers. For the time being, it might be implemented using a multi-criterion decision rule in which the return to the investment is considered along with a number of operational aspects. These may include the imposition of safe minimum standards in environmental protection and limits on the rates of depletion of certain key resources for which substitution options are limited. Also, when natural capital such as mineral reserves or forests are depleted, it will be wise to make certain that other compensating investments are made.

THREATS TO SARD

For thousands of years, agriculture, including forestry and fishing, has provided the food and other materials necessary for human survival and well being. Yet today, as we have seen, there is growing concern about the capacity of agriculture to meet the needs of future generations. The challenge to SARD comes from population growth, the increasing demand for food, the need to reduce poverty, and the threat to agricultural resources of land, water, and living things.

Population Growth

By the year 2000, world population is expected to be about 6.25 thousand million, compared with 3.0 thousand million in 1960. About 80 per cent of the year 2000 population will live in the less developed regions. While the global rate of population growth has slowed to less than 1.7 per cent annually, the absolute increase, of nearly 100 million annually is more than it has ever been. Even if birth rates are brought down relatively quickly, the high proportion of young people in those countries where population growth is still high means that the momentum for growth will continue at least until the middle of the next century.

The rates of population growth are very uneven, with most industrialised countries experiencing close to zero rates of natural increase, and most of the growth coming in the less developed countries. In some parts of the world, as in much of sub-Saharan Africa, population growth has tended to outstrip growth in food production. It is in just these less developed parts of the world that future food production needs to be increased more rapidly in order to improve food security, especially for the poor.

Growing Demand for Food and Fibre

Global demand for food is likely at least to double by 2030. Most of this increase will occur in the less developed countries of Africa, Asia and Latin America. This concentration in part reflects the rapid population growth in these areas. It also reflects, all being well, desirable improvements in the quantity and quality of food consumed per head as incomes rise. In sub-Saharan Africa, to feed the increasing numbers and to reduce presently unacceptable levels of food insecurity require a more than tripling of agricultural production by 2025 from a 1990 base.

In addition to the population growth, changes will occur in the pattern of demand. In many of the newly industrialising countries, and in those less developed countries that have been successful in promoting economic growth, rising incomes are likely to lead to significant increases in demand for meat and other animal products. Such changes in demand for animal protein are unlikely to be fully or even mainly met from livestock grazed on land not suited for cropping. They therefore imply an increased demand for grain to be used as feed for intensively kept livestock and farmed fish. To produce one kilogram of human food as intensively reared animal protein takes several kilograms of grain or other stockfeed. If there is not a compensating increase in grain production, the diversion of these feeds to animal production is likely to drive up prices, reducing the access of the poor to staple foods.

The increased food production to meet the expanded demand will have to come mostly from increased yields. The reason is that there is little new land to be brought into production, and agricultural land is being lost to urban development, degradation and for other reasons. During the 1970s and 1980s, arable and permanent crop area per capita fell from 0.37 ha to 0.27 ha for the world as a whole. The fall in the less developed countries was sharper from 0.27 ha to 0.19 ha. These trends in arable area are likely to continue. They mean that the projected doubling in the demand for food by 2030 will be met only if yields per hectare are more than doubled. At best, this is a daunting challenge. There are some who suggest that it is not attainable, or at least, not without seriously damaging consequences for the environment as a result of the required increases in agricultural intensity based on increased use of chemical inputs.

Challenges to Reduce Poverty

According to IFPRI, over 1.1 thousand million people in the developing world live in absolute poverty, with incomes per person of a dollar a day or less. Moreover, IFPRI estimates that about 800 million of these, or one in five of the population of the developing world, lack the economic and physical access to the food required to lead healthy and productive lives. There is an urgent challenge to agricultural policy makers to find ways of reducing the scale of this deprivation, above all for humanitarian reasons. In addition, failure to tackle the problem quickly and effectively is likely to mean that the rural poor will be driven to migrate to the cities in search of livelihood.

The swelling slums that surround many of the major urban centres in less developed countries present their own threats to the sustainability of social and political systems. Poverty persists because people lack entitlements that they can use to earn an acceptable living. They may have no access to land, or to water to irrigate arid land, or to technologies appropriate to their needs and circumstances, or to markets that offer reasonable prices for what they can produce, or to employment opportunities that offer a reasonable reward for their labour. These problems of lack of entitlement are not easy to solve, yet finding solutions is fundamental to the pursuit of SARD. Unless policies for SARD lead to a reduction in poverty, they cannot be regarded as successful.

At least in some situations, there is a vicious circle linking poverty and unsustainability. The poor may be driven by their very poverty to degrade resources in order simply to survive. Certainly, they are likely to lack the capital and access to credit to invest in resource improvements such as land conservation measures. Moreover, as the resource base declines, at least per person, the pressures on the poor that force them into unsustainable ways may increase.

In such cases, policy makers need to find a means to break the downward spiral. For example, a solution might lie in the provision of a 'food for work' program, with the labour used for conservation measures. Yet too much should not be made of the contribution of the poor to resource degradation, since it is the rich who are the largest consumers of global resources such as energy. Moreover, the kinds of resource degradation caused by the rich (global warming, pesticide run-off) tend to be different from the damage done by the poor (deforestation, erosion). Clearly, attention needs to be given to ways of changing the behaviour of both rich and poor to address these different issues if sustainability is to be attained.

Resource Degradation

Land, Water and Environment

Hard data on degradation of land and water resources on a world scale are lacking. There is much concern, no doubt much of it well justified, about losses due to soil erosion, salinisation or desertification. Yet in a recent review of a number of studies, Crosson and Anderson note that the historical loss of productivity due to land degradation, although not unimportant, has been rather low relative to the productivity increase due to other factors. Even under rather pessimistic assumptions, they estimate that the annual average rate of productivity loss due to land degradation in the less developed countries has been less than 0.2 per cent a year over the past 45 years.

There is also concern about competition for water between agricultural, industrial and urban uses and about declines in water quality for agricultural and other purposes. However, Crosson and Anderson further conclude that the average rate of productivity loss on irrigated land due to degradation of water resources for irrigation has been no more than 0.3 per cent a year over the past 30 years. While the above rates are far from insignificant in the long run, they do suggest that the scope for attaining SARD is likely to lie mainly in expanding the use of knowledge-intensive inputs in agriculture, rather than in giving absolute priority to preventing land and water degradation, or to efforts made with too little regard to costs to reverse past unfavourable trends in quality of land and irrigation water resources.

The shift in technology required to produce more food and fibre from less land has thrown into doubt the capacity of the natural environment to absorb the resulting pollution and contamination. Environmentalists emphasize the role of the environment not only as a source of resources or inputs, such as land and water, used in production, but also as a sink to absorb unwanted by-products of production and consumption, and as a direct source of amenities. They are concerned that this absorptive capacity is being overloaded and will be much more seriously threatened if present methods of agricultural production are greatly intensified. Apart from the perceived risk of a breakdown in the life-sustaining ecosystem itself, there is the lesser but still serious threat to human health as more pollutants from intensification are released into the environment.

The challenge, therefore, is to find more intensive methods of production that impose a lesser burden on the environment. Pretty has argued that there are resource-conserving technologies, local institutional structures and enabling external institutions that meet this challenge. The task for policy makers is to provide the conditions for such positive changes to occur.

Loss of Animal and Plant Species

The development and intensification of agriculture have led to loss of diversity in both domesticated and wild biological species, and the decline is continuing. In the past, humans have used well over 100 000 edible plant species and numerous animal species.

Today, most of the world's population depends on only 12 crop species for their food staples. While the actual degree of concentration is overstated by these figures, in that many of the species used in the past were never very important, many scientists see the implied loss of diversity as worrying. While some loss of biodiversity is unavoidable, and while there are certainly advantages from concentrating on the more productive species, there are grounds for anxiety that the decline may be socially and ecologically excessive. The erosion of the genetic resource base may mean the permanent loss of actually or potentially useful species or traits. Moreover, production systems are exposed to increased risks by relying on a narrow genetic base for domesticated species. SARD clearly requires that the downward trend in biodiversity be arrested.

Challenges to Sustainable Forestry

Deforestation is occurring at a rapid rate in many parts of the world. According to the 1990 Forest Resources Assessment, the annual loss of tropical forests has been about 15.4 million ha, mostly in the less developed countries. Afforestation and reforestation in the tropics have been 1.8 million ha per year, representing only 12 per cent of deforestation. In many of the less developed countries, deforestation is occurring as forest is cleared for farmland and rangeland to support growing human populations. The farming systems introduced on the cleared lands are not always sustainable. While there has been an expansion of the temperate forest resource in the industrialised countries, the demand for forest products is expanding. There is also an increased demand for the services forests provide in protecting the land and water base, providing sustainable livelihoods for forest-dependent communities and conserving ecosystems and biodiversity.

Challenges to Sustainable Fisheries

Despite the rapid development of aquaculture, fishing remains largely a harvesting activity characterised by lack of property and user rights. The result is that many stocks and fishing areas are over-fished. According to one estimate, all the world's 17 main fishing grounds are being fished at or above their sustainable limits. Some 70 per cent of the world's fish stocks have been classified as fully exploited, over-exploited, depleted or recovering from over-exploitation. Moreover, the functioning of these ecosystems and their capacity to recover from over-exploitation is uncertain.

Management of fisheries is not easy and demands cooperation at several levels, from local to international. FAO has developed a Code of Conduct for Responsible Fisheries. The Code places emphasis on the adoption of selective and environmentally safe fishing practices designed to maintain biological diversity, and safeguard aquatic ecosystems.

Challenge in Developing Human Capital

Much of the discussion about SARD focuses on the physical resources used in agriculture. Yet it is clear that substantial increases in productivity will be needed to meet future demand for food. These increases must come from improved technologies, most of which will require large investments in education, research and technology development. The challenge, therefore, is to make sure that investments in these forms of human capital are sustained at the necessary rates. This challenge is critical because it seems that the easiest technological gains may have already been attained, so that a higher rate of investment will be needed in future than in the past to sustain the required rate of productivity increase.

POLICY FRAMEWORK FOR SARD

There needs to be an appropriate framework in place for the formulation and implementation of policies for SARD. Moreover, policy making for SARD needs to be approached accounting for some important features of the task. Elements that are likely to be important include:

— Taking a long-term, global perspective.

— Implementing complementary sectoral and macro-economic policies.

— Developing coordinated and consistent policies within the agricultural sector.

- Having an appropriate legal and institutional structure in place.
- Following an approach in which the inherent uncertainties of planning for SARD are recognised and accommodated.

Taking an adaptive, inter-disciplinary and multi-disciplinary 'systems approach' to policy planning.

Global Perspective

The need to take a long-term view in considering sustainability should not need emphasizing, yet there appears to have been a tendency in recent policy work to concentrate on the short term, perhaps driven by political expediency or short-term difficulties in macroeconomic management. A focus on SARD requires decision makers to take a more far-sighted view of their responsibilities.

The case for a global perspective rests on the point noted earlier that sustainability needs to be assessed at the highest level, and that sustainability of the global system does not require that every sub-system be sustainable. Technological changes in transport, communications and information science are creating a global economy in which goods, services and resources are widely traded. Countries, districts within countries, and even individual businesses and households, are increasingly able to benefit from trading in the progressively more integrated global market, and so need to give less attention to self-sufficiency.

The principle of comparative advantage suggests that freer trade should facilitate increased and sustainable production from the available resources. It implies that countries can import resources that are in short supply, or can export resources that are in relatively abundant supply. There have even been suggestions that countries can export 'pollution' to where it can be better dealt with—a proposition that, not surprisingly, can excite considerable opposition!

The increasingly integrated global market should also mean that prices better reflect the relative global scarcities of resources, goods and services, sending more appropriate signals about these relative scarcities to producers and consumers. This should promote more sustainable behaviour, provided only that governments do not excessively distort domestic prices through inappropriate protectionist or taxation measures. According to Schuh and Archibald:

> The significance of these developments to the sustainable development issue is that this increased integration through international markets creates the potential for large efficiency gains from the international division of labor, increased specialisation in production, and the realisation of comparative advantage. The efficiencies from these processes will release the resources essential for realising sustainable economic development on a scale not realised in the past.

Extending this line of reasoning further, some argue that trade liberalisation will in fact promote sustainable development only if other problems, such as other market failures and inequity, are tackled at the same time. These critics argue that, in a world of rich and poor, it is too easy for the rich to use their trading power to the disadvantage of the poor. Certainly, the process of structural adjustment in a country as trade barriers and other regulations are dismantled often appears to exacerbate rather than cure problems of poverty and inequality, at least in the short to medium term. It may therefore be important for policy makers to give careful thought to the appropriate pace of change and for them to put in place safety net measures for those who will be most disadvantaged.

Harmonising General and Sectoral Policies

Macroeconomic policies can be as important, or more important, than sectoral policies in affecting resource use, income distribution and growth, and hence SARD. It is therefore vital that macro-level and sectoral policies are harmonised. In the less developed countries it has been common to put in place policies that in effect discriminate against agriculture; in the industrialised countries, the reverse is the case. Both types of policies tend to create serious inefficiencies in the use of resources, including agricultural resources, and therefore impede progress towards SARD.

Integrating Action within the Agricultural Sector

It is similarly important to harmonise policies within the agricultural sector. This might not be as simple a task as it may seem, since policies affecting agriculture are not made solely within the agricultural ministry. So, improvements to rural transportation and infrastructure, and to rural health and education, are all vital components of efforts to progress towards SARD. Yet these areas usually fall within the responsibility of other ministries, not the ministry of agriculture. Unless systems are put in place to encourage harmonisation of efforts, and to avoid conflicts between different groups and agencies, efforts may be dissipated.

Building an Appropriate Legal and Institutional Structure

Market failure often occurs when the legal structure is too weak for people to enforce their property rights, or for governments to enforce regulations for the control of resource use. Private property rights usually exist in resources or in commodities. Well-functioning markets can contribute to SARD by helping to direct resources into their most productive uses. Yet markets cannot function unless there are clear and transferable property rights and enforcement of contracts.

When property is held in common, institutions need to be developed, or even created, to encourage sustainable management. For instance, the allocation of irrigation water to competing users requires the establishment and operation of a competent authority. Responsible fisheries require a forum for cooperation among interested parties which has the power to enforce catch limits to prevent over-fishing. In all such cases, the existence of an appropriate legal and institutional structure is clearly necessary for SARD.

Policy-making Accounts for Imperfect Knowledge

Risk, risk aversion and downside risk

The very long-term nature of SARD makes decision making difficult. Errors in policy making or resource management may not come to light until far into the future. Moreover, bad decisions may lead to very bad outcomes. Many resource management choices are characterised by irreversibilities. For instance, an inappropriate land management strategy in an upland area may lead to serious soil erosion and land-slips. Once the soil is lost, it may be impossible to restore the land at any reasonable cost. Similarly, once a plant or animal species becomes extinct, it is lost for ever, no matter how potentially valuable it might have been.

Such a situation might seem to signal a need for very risk-averse policy making in resource management. Yet this is not necessarily the case. In the context of project appraisal, the rejection of options just because they are risky could be a mistake. In so far as the consequences of policy decisions are spread widely and fairly evenly among the population of a country, many policy risks are not important. In many cases, though not in all, it is best to base the choice of policy, at least in principle, on what maximises some measure of expected social value. 'Expected', in this context, means probability-weighted average computed across all possible outcomes, including any very bad outcomes that could occur.

There is an important distinction here between the notion of an expected value averaged across all possibilities, and a measure of social welfare calculated by assuming that 'average', 'best-guess' or 'typical' consequences follow from any policy decision. The former gives some weight to the risk of very bad outcomes which, as discussed, may be considerable when planning for SARD. By contrast, planning on the basis of 'normal' or 'most likely' outcomes means that the risk of such negative consequences is ignored.

The notion of planning to improve expected social welfare does require some judgements about the chances of occurrence of various kinds of consequences. The trouble is that the required probabilities are seldom known. Indeed, because of the long-term nature of planning for SARD, some actual outcomes may not even be conceivable at the time a decision has to be made. Planning in such an uncertain environment seems to call for extra caution. In addition, Little and Mirrlees note some special cases where risk aversion in policy making and planning is appropriate.

A case in point is when the consequences of the decision are closely related to the overall performance of the economy. This may happen either because those consequences are large relative to the size of the economy, and/or they are strongly positively correlated with national income. For a small economy that is heavily dependent on income from fishing, such as some of the small Pacific island nations, for example, policy risks in the management of the fishery resources will obviously be more important than in larger, more diversified economies.

There may also be a case for risk aversion when a policy has the potential to have a substantial effect on the welfare of certain groups, especially when those groups may be particularly vulnerable. Clearly, such a situation may prevail quite often in policy making relating to SARD. If policy fails to secure the sustainability of some agricultural system or rural economy, the local inhabitants may be seriously disadvantaged, even impoverished. In such situations, therefore, there is again a case for a more precautionary approach than basing decisions on expected benefits and costs alone.

Finally, special consideration needs to be given to cases where there are serious irreversibilities. In these situations, basing policy choice on expected values may seriously under-estimate the costs of future opportunities forgone when those opportunities cannot be quantified, and perhaps not even identified, at the time when a decision is taken.

Precautionary Principles and Adaptive Planning

When any policy action or inaction carries with it the possibility of serious and perhaps irreversible resource or environmental degradation, with an associated risk of significant negative impacts on future welfare, actions to avoid or reduce that risk should be carefully considered. Such actions may include:

- postponing a decision to change the allocation or use of resources until more information is obtained about the possibility of serious negative results from the change;
- in a situation where continuation or projection of present practices creates the threat of serious negative consequences, relatively strict resource and environmental protection measures, such as safe minimum standards, may be imposed, at least until more is known.

The first rule might be called the 'look before you leap' principle, and the second might be called the 'better safe than sorry' principle.

In advocating these precautionary principles, it is not being suggested that policy makers should adopt a very risk-averse stance. Many risks to SARD, at least at the level of individual agro-ecosystems, are so small as to be insignificant in the context of policy making and planning at the national level. Precaution is, however, called for when the impact on expected costs and benefits of negative outcomes may be serious, in relation to the 'special cases' set out by Little or Mirrlees. It may also be appropriate when too little information is to hand to know whether the risks could be serious.

Operationally, the principles imply adding to the list of options usually considered by policy makers. Thus, the 'look before you leap' principle implies that, for example, instead of a choice between building a dam to generate electricity or not building it, the option is also explicitly considered of delaying any start to construction for a year or two. In that time, more data can be gathered on environmental impacts, and scope for energy saving that might eliminate the need to increase generating capacity can be explored.

The 'better safe than sorry' principle means that, if there is uncertainty about the safety of a pesticide that is in common use, its use might be restricted until more conclusive evidence is available. Note that, in both cases, some actual or opportunity costs are incurred by acting cautiously, and these costs must be balanced against the achieved reduction in the chance of serious negative consequences.

It might be argued that the very familiarity of the aphorisms used to characterise these precautionary principles shows that they are merely common sense, which indeed they are. Common sense, however, is not always applied in policy making, as many past policy mistakes reveal. Adding a requirement for caution to the agenda established for policy analysis may reduce the frequency of such mistakes.

Systems Approach to Policy-making

Any strategy to address SARD requires a comprehensive perspective that accounts for the interrelationships among the technical, environmental, social, economic, and political aspects of development. It must also provide for explicit recognition of, and understanding of, the social, economic and ecological diversity and uniqueness of each country, region and district.

A systems approach, involving multi-disciplinary teams, offers the best possibility of attaining a broad perspective encompassing these various dimensions. Moreover, there is a need for the strategy, and the choice of specific policy instruments, to be evolutionary. That means that decisions are adjusted or changed as more is learned about the nature of the problems to be addressed and about which solutions work and which do not.

The merits of the systems approach also extend to micro-level analysis for SARD. Selecting appropriate interventions for agriculture often requires understanding the responses of farm households to changes in the external environment. This understanding needs to be built on an appreciation of farm-household decision making. Again, therefore, a systems approach provides a relevant conceptual basis for micro-level analysis for SARD. For example, technological packages for SARD are mostly site specific. They depend on the farming system, size of units, resource stocks and quality, availability of inputs, access to markets, and socio-economic factors. The development and testing of improved technologies to promote SARD need to be tailored to identified main farming systems and validated on farms representative of those systems.

REFERENCES

Boddington, M. 1996, 'Training issues for sustainable agriculture and rural development', in S. A. Breth (ed.), *Integration of Sustainable Agriculture and Rural Development Issues in Agricultural Policy*, Winrock International Institute for Agricultural Development, Morrilton, Arkansas.

Carley, M. 1994, *Policy Management Systems for Sustainable Agriculture and Rural Development*, International Institute for Environment and Development and Food and FAO, Rome.

Chambers, R. and Conway, G. R. 1992, *Sustainable Rural Livelihoods: Practical Concepts for the 21st Century*, Discussion Paper 296, Institute of Development Studies, Falmer, Sussex.

FAO. 1989, 'Sustainable development and natural resources management', Twenty-Fifth Conference, Paper C 89/2 - Sup. 2, Food and Agriculture Organization, Rome.

Pétry, F. 1995*a*, *Sustainability Issues in Agricultural and Rural Development, Vol 1, Trainee's Reader*, Training Materials for Agricultural Planning 38/1, FAO, Rome.

12

Sustainable Agriculture and Nature Conservation

Protected area management, including biodiversity conservation and local community participation (especially of indigenous people) are main discussion themes of the protected areas constituency. Although relationships and alliances between local communities and conservationists are considered, what is less addressed are options for sustainable livelihoods within and around protected areas. Farmers and forest dwellers are the main inhabitants and users of protected areas as well as lands connecting these areas. In protected area categories where agricultural activities are allowed, there is need to consider productive activities which provide livelihoods in an equitable and environmentally-friendly way.

The ability of organic agriculture, ecoforestry and sustainable forest management to build self-generating food systems and connectedness between protected areas is addressed in this paper. Also, farmers' involvement in income-generating activities such as agro-ecotourism, is considered. This paper argues that integrated landscape planning is necessary for the effective management of protected areas and that multi-sectoral policy and planning, including the agriculture, forestry, tourism and environment sectors, have a role to play in such a collaborative resource management. The ultimate objective is to recognize the interdependence between sustainable agriculture and biodiversity conservation and in so doing, promoting options that address food and livelihood needs while protecting the natural heritage.

PROTECTED AREAS IN AGRICULTURAL LANDSCAPES

The aims of protected areas and approaches to their management have recently expanded. Land managers can use protected areas as a tool for a wide range of functions. Among the six IUCN protected area categories, Categories V and VI recognize that maintenance of biodiversity is not always the primary reason for protection and that (management) choices may be determined by cultural values, environmental management, sustainable land use and recreational needs. In particular, these categories are suited for sustainable use by indigenous communities, tourism and small-scale agriculture.

The definition of protected areas as "an area of land and/or sea especially dedicated to the protection and maintenance of biological diversity, and of natural and associated cultural resources, and managed through legal or other effective means" implies that management of protected areas is inclusive of government-sponsored reserves, indigenous communities or private landowners. Control and management by stakeholders concerned have proven to be effective while providing livelihoods to local communities.

The protected area are categorised as follows:

- *Category Ia: Nature reserve.* Strict nature reserve/wilderness protection area managed mainly for science or wilderness protection—an area of land and/or sea possessing some outstanding or representative ecosystems, geological or physical features and/or species, available primarily for scientific research and/or environmental monitoring.
- *Category Ib: Wilderness area.* Protected area managed mainly for wilderness protection—large area of unmodified land or sea, retaining its natural characteristics and influence, without permanent or significant habitation, which is protected and managed to preserve its natural condition.
- *Category II: National park.* Protected area managed mainly for ecosystem protection and recreation—natural area of land and/or sea designated to (a) protect the ecological integrity of one or more ecosystems for present and future generations, (b) exclude exploitation or occupation inimical to the purposes of designation of the area, and (c) provide a foundation for spiritual, scientific, educational, recreational and visitor opportunities, all of which must be environmentally and culturally compatible.
- *Category III: Natural monument.* Protected area managed mainly for conservation of specific natural features—area containing specific natural or natural/cultural feature(s) of outstanding or unique value because of their inherent rarity, representativeness, aesthetic qualities or cultural significance.
- *Category IV: Habitat/species management area.* Protected area managed mainly for conservation through management intervention—area of land and/or sea subject to active intervention for management purposes as to ensure the maintenance of habitats to meet the requirements of specific species.

- *Category V: Protected landscape/seascape.* Protected area managed mainly for landscape/seascape conservation and recreation area of land, with coast and sea as appropriate, where the interaction of people and nature over time has produced an area of distinct character with significant aesthetic, ecological and/or cultural value, and often with high biological diversity.
- *Category VI: Managed resource protected area.* Protected area managed mainly for the sustainable use of natural ecosystems—an area containing predominantly unmodified natural systems, managed to ensure long-term protection and maintenance of biological diversity, while providing at the same time a sustainable flow of natural products and services to meet community needs.

The focus of protected area management has shifted away from individual protected areas towards protected area networks, as part of wider landscape. The effectiveness of protected area management depends on this network as well as on the type of land use in between these areas: islands of biodiversity in an otherwise degraded landscape will not fulfil protection objectives. Opportunities for conservation of biodiversity outside protected areas include links between protected areas through corridors and provision of environments free of pollutants, especially but not exclusively, along migration pathways.

Protected areas cover only about 10 percent of the earth's cover: 45 percent of the world's protected areas maintain 30 percent or more of their land in agriculture. Most wild plants and animals live outside protected areas, often in agriculture-dominated landscapes: about 30 percent of the global land surface is occupied by crop and managed pasture lands. People live inside and near protected areas and use land, plants and animals to meet their basic livelihood needs. Depending on the type of management practices, farmers and foresters will have varying impacts on habitat quality, which can increase or decrease pressure on biodiversity within and around protected areas.

PRESSURE OF AGRICULTURE ON BIODIVERSITY

Over the last century, population, market pressures and the development of new agricultural technologies have encouraged patterns of agricultural development tending towards agricultural intensification (i.e. increasing scales of monoculture production, intensive mechanical tillage, irrigation, and the use of synthetic fertilizer, pest control agents and a restricted diversity of crop and livestock varieties), often leading to natural resources degradation. The growing food demand by a wealthier and larger global population is expected to induce further encroachment of agriculture on unmodified ecosystems (10 billion hectares by 2050), with inevitable negative impact on biodiversity.

The majority of the human population increase is expected to take place in the biodiversity-rich developing countries of the tropics (e.g. the Caribbean, the Philippines,

Sri Lanka and the Western Ghats of India), where 19 out of 21 regions of concentrated biodiversity ("biodiversity hot-spots") and human population in these areas is increasing faster than anywhere else). These areas of high population growth (many of which lie adjacent to protected areas) are also experiencing rapid changes towards urbanization where demand for agricultural products is expected to increase as income levels in these areas rise. The anticipated result of such demographic changes is that increased production pressures will be placed on both the wild lands and the agricultural production systems in and around protected areas.

The simplification of agro-ecosystems to monoculture production and the removal of non-crop vegetation from the farm unit (e.g. hedgerows, shelter belts and field margins) has contributed to the homogeneity of agricultural landscapes by reducing botanical and structural variation, resulting in both a reduced capacity of agricultural areas to serve as habitat for wild species as well as to effectively internally regulate populations of pests and disease causing organisms which affect crop productivity. This has resulted in a widespread decline in farm species abundance and diversity across many taxonomic groupings, including high rates of wildlife mortality and reduced reproductive success of many species. This loss of biodiversity has also resulted in a reduced capacity of agro-ecosystems to perform many essential ecosystem functions such as purification of water, internal regulation of pests and diseases, carbon sequestration, and degradation of toxic compounds.

Elevated nitrogen and phosphorus levels in aquatic ecosystems have led to extensive eutrophication and degradation of freshwater and marine ecosystems in many areas where agriculture is concentrated. Synthetically compounded nitrogen fertilizer poses multiple risks to both wildlife populations and human health. Dissolved nitrate levels of 2 ppm or greater are known to interfere with normal development of amphibians with levels above 10 ppm known to be lethal.

The use of pesticides (i.e. herbicides, fungicides, rodenticides and insecticides) poses both known and unknown risks to biodiversity, impacting wildlife on many different levels, from direct to indirect lethality to non-lethal but severely debilitating effects. Each of these impacts has the potential to interfere with the reproductive success of wildlife and further reduce the habitat quality and biodiversity of agricultural and surrounding ecosystems. It is estimated that 70-90 percent of ground applied pesticides and 25-50 percent of aerially applied pesticide reach their target. The remaining amount is released into surrounding ecosystems and enters the food chain, affecting animal populations at every trophic level.

Over 672 million birds are exposed to pesticides each year in California alone with an estimated 10 percent of these animals dying from this exposure. Birds exposed to sublethal doses of pesticides are often afflicted with chronic symptoms that affect their behaviour and reproductive success. Pesticides are also known to negatively affect insect pest-predator population dynamics in agro-ecosystems and to disproportionately effect insect predator populations, resulting in pest population resurgences and the development of genetic resistance of pests to pesticides.

In addition, endocrine-disrupting compounds found in many pesticides still in use pose an additional and unknown long-term risk to wild biodiversity. Significant evidence of endocrine disruption from pesticide exposure has been documented for many different taxonomic groups including: birds, reptiles, fish, snails and oysters resulting in adverse effects to growth, development, or reproduction.

Recent studies have also provided evidence of the impacts and risks to agro-ecosystems and wild biodiversity from genetically engineered crops. Transgenic crops pose a suite of ecological risks to native and cultivated ecosystems through: the spread of transgenes to related wild types via crop-weed hybridization; reduction of the fitness of non-target organisms; the evolution of resistance of insect pests to pesticide producing crops; soil accumulation of the insecticides produced by transgenic crops; unanticipated effects on non-target herbivorous insects; and the creation of new pathogenic organisms via horizontal gene transfer and recombination.

Considering the above processes, it appears that agriculture management will have a considerable impact on the structure, composition and quality of the landscape that predominate in and around protected areas. The simplification and pollution of agro-ecosystems must be avoided or countered by adopting chemical-free and diversified agricultural systems to reverse the decline in species and habitat diversity in and around protected areas.

AGRICULTURE DEVELOPMENT AND BIODIVERSITY CONSERVATION

Due to the predominant agriculture's negative impact on biodiversity, conservation groups and protected area managers have historically viewed agricultural activities as being in conflict with stated conservation goals. This view has often led to attempts by protected area managers at excluding agricultural and other productive activities from protected areas. Similarly, community members in and around protected areas have often viewed measures to conserve biodiversity (e.g. land takings or land and/or water use restrictions) as a threat to personal freedom, livelihoods and the economic viability of their agricultural enterprises. The imposition of such restrictions has often led to real conflicts between protected area managers and farmers.

This perceived and actual antagonism has been compounded by the historic strategies of protected area management that have tended to marginalize the economic, social and political interests of community members around protected areas by excluding them from land management decision-making processes that effect their lives.

Additionally, strategies of excluding agriculture from protected areas have often fallen short of conservation objectives as many species require human-induced disturbances created by agriculture and other land use activities. On the other hand, agriculturalists in and around protected areas are often directly impacted by wildlife. The US Department of Agriculture estimates that annual losses from wildlife damage to crops are 1 billion dollars each year. In developing nations the losses of crops to wildlife are often even more significant and potentially life threatening, with large game animals disturbing fields of staple crop and small rodents causing considerable damage to post-harvest staple goods.

Recently, agriculturalists have become more aware of the value of the biodiversity "input" for agriculture. The ecological functions of diverse ecosystems (such as balanced predation, pollination, nutrient cycling, degradation of toxic compounds, carbon sequestration) are today recognized to be central to sustainable food production. Moving away from simplified agricultural systems offers opportunities to produce food while enhancing natural landscapes.

After decades of struggle, only recently have conservation biologists, protected area managers, agriculturalists and policy makers begun to recognize the biodiversity contribution and potential of agricultural landscapes and started working together in developing strategies to effectively integrate these two land-use activities which will increasingly occupy the same areas of land. However, much research remains to be done in developing agro-ecosystems that serve both human needs and biodiversity.

BENEFITS OF ORGANIC AGRICULTURE

The Convention on Biological Diversity "encourages the development of technologies and farming practices that not only increase productivity, but also arrest degradation as well as reclaim, rehabilitate, restore and enhance biological diversity and monitor adverse effects on sustainable agricultural diversity. These could include, *inter alia*, organic farming, integrated pest management, biological control, no-till agriculture, multi-cropping, intercropping, crop rotation and agricultural forestry".

While several agricultural approaches make sustainability claims, organic agriculture is the only well-defined agricultural management system, including recommended and restricted practices that aim at environmental protection and food production. The decades-long implementation of organic agriculture, including inspection and certification to ensure compliance, as well as the steady growth of organic food sales on the global market, offer a living example of a viable system that reconciles conservation and production needs.

The Codex Alimentarius Commission defines organic agriculture as "a holistic production management system which promotes and enhances agro-ecosystem health, including biodiversity, biological cycles, and soil biological activity. It emphasises the use of management practices in preference to the use of off-farm inputs, taking into account that regional conditions require locally adapted systems. This is accomplished by using, where possible, agronomic, biological, and mechanical methods, as opposed to using synthetic materials, to fulfil any specific function within the system".

All definitions of organic agriculture converge in recognizing that this food system is based on ecological principles and that it relies on using adaptive management for building long-term farm productivity. As such, the principles behind organic agriculture are in line with the Ecosystem Approach advocated by the CBD and its application focuses on functional relationships and processes within ecosystems; enhances benefit sharing; uses adaptive management practices; carries management actions at all level, and where a legislative framework exists, it ensures intersectoral cooperation.

For practical purposes, organic agriculture guidelines and norms specify minimal requirements to label foods as "organic". These requirements contain mandatory and voluntary practices for organic management. Therefore, throughout the world, the contribution of organic agriculture to biodiversity has different levels of achievements, depending on whether the mandatory requirements only are implemented to obtain certification or whether the broader voluntary requirements are also adopted. As such, biodiversity achievements of organic farms range from pollution reduction (of systems based on agro-chemical substitution) to fully-fledged holistic systems that harbour semi-natural habitats and a wide range of biodiversity.

Generally, farm habitat quality in organic farms is enhanced through practices that build below- and above-ground diversity, with a view to establish ecological processes that serve agricultural production. It must be understood that no single organic agriculture practice alone will sufficiently serve to enhance biodiversity but it is the result of the suite of practices, used in concert, that have the net effect of increasing farm habitat heterogeneity and thus, biodiversity.

Reduction of Agricultural Pollutants

Building of soil fertility and avoiding the use of synthetic agricultural inputs (including fertilizers, pesticides and genetically-modified organisms) is common to all organic agriculture norms. In order to ensure the agro-ecosystem productivity, a set of practices are adopted. The main strategies and their benefits on biodiversity are outlined below.

Biological Control

Pest management in organic systems is achieved through a number of general and site-specific physical, biological and accepted biological pest control methods. The primary strategy for pest management in organic agriculture is preventative, where the use of sound soil fertility management serves to improve the physical, chemical and biological properties of soils, leading to elevated soil fertility and optimal conditions for plant growth.

Crops grown in healthy soils tend to be more resistant and resilient to pest and pathogens and require little to no applications of pest control materials. The reduction or absence of synthetic soil fumigants, insecticides, herbicides, fungicides, etc. serves to prevent exposure of agricultural and surrounding ecosystems to the toxic effects of pesticide pollution. Because pesticides have a direct suppressive effect on many organisms within the agro-ecosystem, the elimination of pesticide use removes a major obstacle to the diversification process of agro-ecosystems. In protected area landscapes, the avoidance of pesticide use by organic farmers is conducive to wildlife conservation.

Use of Natural Soil Amendments

The rational use of soil amendments is a critical aspect of organic system management. The use of soil analysis and nutrient budgets matches agricultural inputs with the nutrient demands of crops in order to make more efficient use of available resources and avoid the pollution problems resulting from the over application of agricultural nutrients. The use of naturally occurring mineral soil amendments (e.g. rock phosphate, sulphate of potash) serves to supply essential plant nutrients while reducing nutrient leaching and/or runoff.

Compost is used to improve and maintain soil organic matter levels and, when combined with an appropriate legume/grass cover crop and the incorporation of crop residues, it helps support larger and more diverse populations of soil organisms. Higher soil biodiversity has shown to increase the rate of nutrient cycling, improve soil aggregation and aggregate stability and improve the disease suppression of agricultural soils.

Use of Adapted Genetic Resources

While there are no, as yet, specific requirements for the use of genetic resources in organic certification systems, the restriction on use of synthetic inputs and genetically-modified organisms leads organic farmers to use and regenerate locally adapted landraces and genotypes resistant to climatic stress and diseases. It is also economically advantageous for organic farmers to use traditional varieties that are productive under low-input conditions, that can survive adversities and that can grow in marginal environments. The organic market often valorises endangered under-utilized and threatened species, including varieties and breeds that have specific nutritional characteristics (e.g. gluten-free crops). Research on selection, improvement and distribution of open-pollinated varieties is at early stage of development.

Polycultures

Polyculture is the intentional cropping of multiple species within a farm, in a planned spatial sequence. Intercropping is a direct way of increasing the diversity of an agro-ecosystem by introducing much more structural and chemical variation to smaller areas of the agricultural landscape and thereby encourages more complex interactions between crops and other organisms.

Strip cropping, where strips of one crop are planted next to strips of another is a less management intensive form of multi-cropping. When compared to monocultures, polycultures have consistently shown lower populations of pest insects and weed populations. Increases in the variety of food sources, the creation of micro-habitats as well as the increased difficulty of pest populations within polycultures to locate dispersed patches of crop hosts serve to limit the growth of pest populations and stabilize predator-prey and parasitoids-host population dynamics.

Crop Rotation

Crop rotation is the intentional planting of crops in different areas within a farm, in a planned temporal sequence. Crop rotations serve to provide new above- and below-ground habitats, as each new crop has a distinct chemical and biological make-up, introducing new vegetation types to the landscape and eventually, crop residues to the soil ecosystem. Different crop residues promote or inhibit different soil organisms which may have inhibitory or growth promoting effects to subsequent crops or pests. By interrupting the continuous presence of a crop host, crop rotation serves to break the build-up phase in the cycles of weeds, insects and diseases, thus eliminating the need for pesticide application. Fallow periods, where ground is left uncultivated for an extended period of time (a few months to one year), allow a limited amount of secondary succession to advance and hence, the recovery of the diversity of both terrestrial and below-ground species.

Cover Cropping

Cover cropping is a type of non-crop cover grown specifically for soil improvement purposes. Both annual and perennial cover crops may be used, with varying benefits to above- and below-ground biodiversity. Cover crops may provide a physical temporary habitat for many different species of ground-nesting birds and small mammals as well as nectar and pollen sources for many species of insects. The habitat value of cover crops varies by species and variety; therefore, cover crops must be carefully selected to meet specific management objectives.

Cover crops root system improves water penetration and prevents soil erosion. Cereal cover crops planted as an over-wintering rotation may also capture plant available nitrogen, thus preventing nutrient leaching into sensitive water ways. Cereal and legume cover crop mixes can be an important source of organic matter when incorporated into the soil. The use of perennial cover crops in orchards and vineyards is an effective means of enhancing the biodiversity and productive capacity of cropping systems and avoiding the labour and environmental risks associated with herbicide use.

In Mediterranean climates, economic and agronomic advantages may be gained through the use of native perennial grass species in vineyards having a summer dormant growth cycle. As a result, they provide the advantages of a perennial cover crop without the disadvantage of excessive competition or water use.

Minimum Tillage

Although reduced tillage is not a required practice in certified organic agriculture, it is a practice that is consistent with the objectives of organic system management and encouraged by most certifying programmes. The minimization of tillage will often lead to increased earthworm abundance and activity, increased populations and diversity of decomposer organisms and an associated increase in the organic matter content and aggregate stability of soils.

When compared to conventionally tilled soil, low and no-till systems exhibit improved nutrient and water holding capacities, improved nutrient cycling and more desirable physical properties. The enhanced biodiversity of these systems lend themselves to improved cropping performance and greater resistance and resilience to various kinds of disturbances. Farming systems using no and low-till approaches are also less likely to incur soil erosion and require less inputs of energy.

Impact of Organic Agriculture on Biodiversity

Numerous studies have provided evidence of the positive role organic farming plays in both above and below-ground biodiversity, reduction of agricultural pollutants and the preservation and restoration of on-farm biodiversity. Long-term comparisons with conventional agriculture systems demonstrated that organically farmed soils showed higher biological activity (30-100 percent) and higher total mass of soil micro-organisms (30-40 percent). Nitrate leaching rates on organic farms were shown to be significantly lower (40-64 percent) and energy use to be more efficient (30-50 percent) on a per hectare basis.

Studies taking place between 1988-2001, comparing conventional and organic agricultural practices in both the US and UK have repeatedly shown higher levels of wild biological diversity (e.g. birds, arthropods, weedy vegetation and soil organisms) in organically managed farms. The influence of organic agriculture and landscape diversity was further illustrated in over 30 separate studies contrasting organic and conventional farms in the UK between 1983 and 2000. The findings showed that organic systems consistently had higher levels of: wild plants (5 times more biomass and 57 percent more species); arthropods, non-pest lepidoptera and spiders; and birds. In particular, 25 percent more birds were found at field edges, with 44 percent more in-field birds in the fall-winter, and higher numbers of breeding pairs of rare bird species.

Creation of Habitats

Although biodiversity conservation is implicit in most national and international organic standards, explicit biodiversity conservation requirements for non-agricultural purposes remain under-developed at this time. In recognition of the need to establish standards for biodiversity conservation in organic agriculture, the 2002 revision of the IFOAM.

Basic Standards for Organic Production and Processing outlines a set of recommended practices that may serve as a foundation for establishing certification standards for biodiversity conservation in agriculture. These standards' section on "Organic Ecosystems" requires measures for maintaining and improving landscape and enhancing biodiversity quality as well as prohibition on clearing of primary ecosystems.

General recommendations are made on the maintenance of an appropriate portion of the farm as wildlife refuge habitat through, *inter alia*: the establishment of hedges, hedgerows and ecologically diversified field margins as well as areas with ruderal flora and wildlife corridors that provide linkages and connectivity to native habitats.

Agro-ecosystem heterogeneity (and therefore biodiversity of the landscape) may be increased through the conservation, management and/or restoration of various on-farm landscape features such as soil, water and native non-crop vegetation. The techniques outlined below require varying degrees of modifications of current agricultural practices.

Not all organic systems implement habitat enhancement practices and there is scope for improving organic standards for biodiversity conservation. Additional research in restoration ecology will be necessary to make specific recommendations on how to best restore native vegetation to altered landscapes in particular areas. Also, the cost implications of restorative measures need to be compensated for the effective uptake of truly holistic systems, especially in the neighbourhood of protected areas.

Minor Farm Habitat Enhancement Strategies

A number of on-farm management techniques may be used to protect or enhance on-farm habitat for wild biodiversity. Leaving natural snags and the erecting of artificial perches or houses for native song birds, raptors and bats may provide on-farm roosting, nesting sites and refuges to support populations of wildlife. Where mechanical equipment is used for on-farm vegetation management, harvest or clearing of land, the postponing of such operations until after ground-nesting fledge lings have emerged may significantly decrease nesting losses.

Development of wetlands and multi-purpose farm ponds. The extensive loss, worldwide, of natural wetlands makes agricultural wetlands most important to wildlife. In particular, flooded rice fields have an enormous significance in creating temporary wetlands for reproduction and feeding of migrant birds and other wildlife (e.g. reptiles, amphibian). These important agriculture wetland functions are best addressed by organic management that avoids using synthetic pesticides and fertilizers.

In other environments, the development of farm ponds may serve to attract diverse wildlife and may be an economical source of water for small farmers. Planted with native aquatic vegetation, farm ponds provide food and cover, resting and breeding areas for wildlife. Migratory waterfowl may be attracted to these ponds for food resources and may become occupied year around with salamanders, turtles, shorebirds and certain species of ducks. Wide ranging animals may also use small farm ponds for watering holes. In agricultural areas reliant on furrow irrigation, tail water pond systems are an effective way of recycling irrigation water, trap and filter sediments and provide year around water habitat for wildlife.

Leaving strips of unharvested Crops for Wildlife

Leaving some unharvested portion of fields for wildlife forage has proven a highly effective means of maintaining or enhancing population levels of many species. Food plots provide over winter food for wildlife which is particularly useful during times of stress or in areas when habitat modification has been significant. Location, size and spacing are several factors to consider when planning food plots. Although plot size and shape may vary according to the type of planting and the intended wildlife use, it is commonly recommended that wildlife lots be scattered over the entire property if possible.

Corn is the most common forage plant for wildlife, but annual rye, millet and buckwheat are also beneficial for both mammals and waterfowl. In temperate climates, perennial crops such as clover, alfalfa and other legumes can be planted to provide food for turkeys, songbirds, rabbits and deer in the summer. In addition, sunflower beds along field edges provide more food for birds and small animals.

Constraints of this method of wildlife habitat enhancement may include unintended attraction of animals to production areas, leading to crop losses and artificially elevating the carrying capacity of lands resulting in population declines upon discontinuance of practice.

Native Hedgerows

Intentional plantings of complex assemblages of native trees and herbaceous and woody perennials along field margins and areas of marginal or highly erosive soils is an effective means of enhancing biological diversity and the production capacity of agro-ecosystems. Hedgerows introduce greater micro-habitat diversity to agriculture lands and can reduce the velocity of winds and runoff when planted along field margins. As a result, hedgerows reduce soil degradation, increase crop yields by conserving soil moisture, increase livestock productivity by buffering extreme temperature variations. The functional role of hedgerows is being researched to enhance biological control of agricultural pests through specific hedgerow species assemblages.

Complex hedgerows provide shelter and food (e.g. a diversity of pollen, nectar and seeds) which serve to elevate the carrying capacity for wildlife in an otherwise simplified agricultural landscape. They also serve as a corridor for the movement of animals between intact patches of habitat that have been interrupted by agricultural development. When hedgerows are planted between natural ecosystems and agricultural lands, they can also help to reduce the potential impacts of agriculture on native plant and animal communities by modifying winds, moisture levels, temperature levels and solar radiation.

Vegetative Buffer strips

Vegetative buffer strips are mixed native vegetation of various widths that are planted along field margins, irrigation and drainage ditches, tail water ponds or between agricultural areas and sensitive native habitats to serve both an environmental protection and biodiversity enhancement function. Being climate adapted, regionally specific native grasses and woody perennials will need only minimal care until they become established (many perennial native grasses and shrubs will also re-seed themselves).

Grass species often have extensive root systems that may serve to prevent erosion and catch sediment and nutrients, thereby filtering runoff and improving water quality. Such forms of vegetation management serve to increase the complexity of vegetation in agricultural landscapes, providing diverse pollen and nectar and micro-habitat for native species of insects, birds and small mammals.

In organic agriculture, vegetative buffer strips also function to protect from the negative impacts of neighbouring conventional farms, as they catch sediment and excess nutrients or agricultural chemicals. Vegetative buffer strips can also serve to reduce the spread of fire from the agricultural areas to the native vegetation. Planting complexes of regionally specific native perennial grasses and other herbaceous and wood perennial species in buffer strips can be an effective and inexpensive way to address the common management problem of weed control in these areas.

Riparian Corridors

Due to the fragmentation of habitat common in agricultural areas, the maintenance or re-establishment of corridors between intact habitat remnants is an important component of enhancing the habitat value of agricultural areas. Riparian vegetation connects terrestrial and aquatic ecosystems creating extensive ecotone areas and serve as corridors for the movement of plants and animals between areas of undisturbed habitats. Riparian corridors are considered to be one of the most diverse and complex terrestrial habitats on earth.

Riparian corridors play a critical role in filtering sediments, providing organic matter to aquatic organisms, regulating water temperatures and dissolved oxygen levels, sequestering excess nutrients and absorbing surface runoff and preventing or reducing flood damage. Loss of riparian vegetation can significantly alter the function of these systems leading to loss of migration corridors, colonization of invasive exotic plant and animal species, stream bank erosion, alteration of streambed habitat, reduced ability to assimilate chemical and soil losses from agriculture lands and associated degradation of water quality.

Maintaining the quality of the water mitigates the need for watercourse clean-up, and encourages wildlife and aquatic life. The UK Soil Association standards for riparian buffer zones for water quality protection recommend a minimum of 2 metres buffer of riparian vegetation on each side of a stream. In Northern California agricultural areas, a minimum of 30 metres on both sides of the stream is often recommended to provide significant ecological values such as wildlife corridors and biological control of pests.

Buffer Zones

Buffer zones link protected areas to one another through the preservation of habitat. By definition, these areas are intended to remain under wild cover or be managed in such a way as to ensure that human land use is compatible with the maintenance of a high degree of biological connectivity. Buffer zones are used in protected areas to delineate areas where human activities such as grazing, non-extractive agroforestry, under-story cropping, hunting and gathering may occur.

Depending on the species selected and how the area is managed, buffer zones may serve to reduce the pressures placed on intact native vegetation for the production of foods and fuels for human consumption. Buffer zones play an essential ecosystem function by allowing for the dispersal of plants and the movement and migration of animals between intact habitat types. To be effective, buffer zones must offer suitable habitat to wildlife and therefore the preservation of such environments free of pollutants along such pathways is of critical importance.

Under-story Cropping

Under-story cropping is one the most effective ways to conserve native habitats and biological diversity of tropical forest areas that are used for agricultural production. Many traditional cropping systems integrate crops into the under-story vegetation of thinned primary or old secondary forests for the production of coffee, cacao, yerba mate and some tea varieties.

Those traditional systems, which minimally modified the native habitat, are currently replaced by intentionally planted polyculture systems composed of native and non-native species that serve various ecologic, economic and cultural functions. In the last decade, the biodiversity value of shade cropping and its role in the sustainable management of protected area landscapes has emerged and a return to traditional polycultures is being achieved through organic agriculture and other biodiversity-friendly approaches. Under-story cultivations present enormous opportunities for the conservation and enhancement of protected areas and buffer zones while securing food and livelihoods for millions of people living in tropical forests.

COMPLEMENTARY STRATEGIES

Landscape Planning to Reduce Habitat Fragmentation

In addition to on-farm management through organic and habitat enhancement practices, the success of biodiversity conservation in protected areas will depend upon how well individual farms are integrated into the wider landscape. Through proper planning at a landscape level, natural and managed areas may be suitably integrated into the landscape in order to reconcile human activities with the goals of biodiversity conservation. In order to best integrate the multiplicity of objectives within protected areas, land managers need effective planning tools to properly manage multi-use areas for the benefit of all interests.

Landscape ecology is the study of environmental factors and interactions at a scale that encompasses more than one ecosystem at a time. Because landscape ecology helps to understand how different parts of the landscape mosaic are formed and how they interact, it may provide a basis for the effective management of agricultural landscapes within protected areas. Landscape planning provides a tool for the integration of agro-ecosystems with the broader landscape in a manner that minimizes habitat fragmentation and that confer the beneficial aspects of biodiversity to agro-ecosystems.

The tools commonly used in landscape planning are mapping, and where available, aerial photography and geographical information system analysis. Such tools provide a means through which historical, current and proposed landscape features and land-use activities may be contrasted to specific conservation goals. With this information, recommendations for changes in land-use practices and agro-ecosystem design may be made that serve to best integrate agricultural ecosystems and other human land uses into the broader landscape. In doing so, impact to critical habitat and sensitive areas may be mitigating or minimized and greater coherence between on-farm habitat and the surrounding landscape may be made. Such management tools may also serve in monitoring the effects of specific land-use practices in order to gauge the effectiveness of interventions.

Landscape ecology and planning may be used at various scales. At a larger scale of analysis, landscape planning may be used to determine the most appropriated placement of agriculture and other land-use activities when attempting to reconcile human livelihoods and habitat needs for biodiversity conservation. By looking at the landscape of which individual agriculture ecosystems are a part, key ecological features (e.g. surrounding plant associations/vegetation patterns) may be identified for use to determining appropriate on-farm habitat restoration efforts that may best serve to re-establish continuity of habitat types and thereby reduce habitat fragmentation. If such ecological features are managed successfully between a number of contiguous farms and wild surroundings, a landscape that is more compatible with the needs of a more diverse number of plants and animals may result.

Agro-ecotourism

Several CBD references mention the link between ecotourism and agriculture: "ecotourism provides for full and effective participation and viable income-generating opportunities for indigenous and local communities". "In order to contribute to the sustainable use of diversity through tourism, there is a need to implement flexible mix of instruments such as integrated planning ... ecolabelling ..." "In some areas, low-input and small-scale agricultural activities that result in both an attractive environment and the maintenance of high levels of biological diversity can offer an opportunity for tourism. Sale of products derived from sustainably harvested natural resources may also provide significant opportunities for income-generation and employment".

Ecotourism. Under the right circumstances, ecotourism has proven to be one of the most effective means to finance biodiversity conservation. In most rich biodiversity areas, actual revenue flows for ecotourism are better than non-timber forest products and biopharmacy, and comparable only to agroforestry (European Preparatory Conference 2002). Because the dominating land use in protected areas and buffer zones is agriculture and forestry, ecotourism is an opportunity for the creation of additional income to farmers/foresters and to generate financial means for the management of protected areas, especially where governmental park management agencies have little resources.

Agrotourism. The symbiotic relationship between tourism and agriculture that can be found in agrotourism (i.e. holidays on farmland) is a key element of an environmentally and socially responsible tourism in rural areas. Rural hospitality offers new employment and income generating opportunities for rural populations, including agrotourism as expression and cultural exchange of agricultural practices, artistic heritage and craftsmanship and culinary traditions. Agrotourism may take several forms: holiday farms, farmhouse bed-and-breakfast, farm camping, mountain resorts, equestrian centres and other forms of rural accommodations. Such facilities are an innovative payment system for environmental services generated on and around agricultural lands.

Agro-ecotourism. While ecotourism is nature-based and agrotourism is farm-based, agro-ecotourism is a combination of both. The rural landscape, usually a combination of wild and agro-ecosystems, is the most important resource for tourism development. It is obvious that a diversified agricultural landscape, with semi-natural habitats, has a greater aesthetic and recreational potential over uniform, degraded and/or polluted agricultural areas.

In Europe, agri-environmental policies often promoted organic agricultural activities as a most effective means for landscape conservation: for example, the European Union Life Environment project run by the French Federation of Parks and Reserves adopted extensive animal husbandry to prevent the negative impacts of unmanaged forests on some botanical meadow species and to promote a landscape quality attractive to tourists.

Examples from the Alpine Region showed that agriculture (e.g. in Carinthia, Austria) maintained an ecological value much more attractive to tourists than areas where agriculture activities were extremely reduced. Tropical countries that harbour extraordinary biodiversity have an untapped potential for generating tourism business around biodiversity-rich farms. For example, shade cacao and coffee farms have a higher biodiversity than forest habitats: families could receive money for visitors access to their land for bird-watching or could be actively involved in the agro-ecotour.

Eco-organic tourism. When agro-ecotourism evolves around an organic farm, it is referred to as eco-organic tourism. The valorization of specific elements of the agro-ecosystem landscape offers an additional economic resource for environmental protection. Conversion to organic management in agricultural areas and the development of connected activities such as tourism are increasing. When farms are organically-managed, they increase the motivation for tourists' visits. New tourist expectations have enhanced the quality of the supply such as diversified farm landscape, environmentally-sound farm-house architecture and local/typical gastronomy.

In many industrial countries, protected area landscapes including farmland experience land abandonment. In this context, small-scale agriculture can become economically viable if quality products could be marketed and income is supplemented by tourism activities, especially in areas rich of biodiversity and history. While the establishment, by local authorities, of land protection systems was historically opposed by farmers unions, there is a recent growing awareness that a reserve/park increases "green" tourism opportunities and that visitors are increasingly sensitive to quality, both in ecological and gastronomic terms. This trend favours organic farmers because they can easily meet these new tourists' demands.

Different studies in the European Union demonstrated the ability of organic agriculture to create attractive landscapes. Eco-organic farm practices and activities which benefit the environment while rewarding farmers include: accommodation in buildings renovated/built according to ecological architecture (natural materials, bioclimatic criteria, landscape planning); on-farm consumption or selling of organic foods and beverages; educational programmes and training (e.g. organic gardening, compost making, wild herbs collection and drying, traditional food and beverage processing), and sensitising guests on rational use of natural resources.

Ecolabels are important marketing instruments for agro-ecotourism in general, because price premiums encourage farmers' commitment to the conservation and maintenance of biodiversity. The most well-known forms of certification include organic farming operations, organic and specialty foods (i.e. geographical denomination of origin) and forest stewardship products. Organic certification of farmhouse structure and facilities is less known but where implemented (e.g. Austria, Italy), it attracts more environmentally-conscious tourists. Requirements for such farms include: organic agriculture production; naturalistic and didactic activities; natural resources tutorship (e.g. at least 5 percent of the farm must be devoted to ecological infrastructure and at least 40 botanical local species must be present in the infrastructure); recreational green areas; ecological buildings (with respect to construction materials and cleaning agents used, energy saving and waste management, and prevention of air pollution); tourist offers (both on-farm and in neighbouring natural reserves); gastronomic offers (organic, seasonal and local); and sustainable transportation facilities.

The dependence of organic farmers on neighbouring protected areas to attract tourists to their farming enterprise and rural hospitality qualify them as best allies for the sustainable management of protected areas and buffer zones. Furthermore, eco-organic tourism offers opportunities for rural economies and sustainable tourism.

Ecoforestry

The importance of ecoforestry and Sustainable Forest Management (SFM) to protected area management is that these approaches to forestry provide opportunities to generate financial and livelihood benefits to local communities which give them an incentive to support conservation initiatives such as protected areas. Ecoforestry has its roots in environmentalism, whereas SFM originates in production forestry.

According to the Ecoforestry Institute of British Columbia in Canada, ecoforestry is a long-term ecologically sustainable and economically sound alternative to current conventional forest management. It is predicated on maintaining the "natural capital" of the forest ecosystem, while allowing a wide range of values and benefits to be derived from the "interest" of the forest.

Nature knows best how to manage forests. By working within the limits of natural processes, we can be sustained in perpetuity. Ecoforestry strives to conserve the structure, function and composition of the forest.

Self-sufficient and stable human communities can grow strong from a sense of place, recognition of interdependence and respect for the forest. Some key values of ecoforestry are ecological integrity, community vitality and economic opportunity. Its methods are specific to bioregions and forest ecosystems and are evolving as we learn more. Ecoforestry favours value-added manufacturing and local jobs by providing a continuing, diverse and local supply of forest products.

Ecoforestry is a low-impact approach to forest management that maintains a fully functioning forest within the natural historic range of spatial and temporal variability. Its practices favour native tree and plant species which provide for the needs of wildlife and their habitats.

Examples of the basic ecoforestry principles and practices are:

- observing the structure, function, composition and natural changes of forest ecosystems, learning from these and using management practices that mimic them;
- preserving the natural diversity of ages, heights and species of trees;
- protecting wildlife and their habitats;
- managing for logs as just one possible product among many non-timber forest products;
- focusing on what should be left after harvesting (in order to keep the ecosystem functioning) rather than focusing on what one takes;
- using low-impact logging systems;
- the volume of trees removed is less than the forests growth rate in order to provide for forest structures such as standing and downed dead trees;
- promoting natural regeneration;

- appreciating all the other forest values (aesthetic, spiritual, genetic, recreational, protective ...) at least as much as the monetary value of marketable products;
- the precautionary principle: when in doubt, as to whether a potential action in the forest may be harmful to the ecosystem or not, don't do it!

The archetypal ecoforester is Merve Wilkinson of Vancouver Island off the coast of British Columbia, who describes his practice in words which exemplify the values and the benefits of ecoforestry: "I'm just completing the 13th cut since 1945 on my 137-acre mixed-species forest. Years ago, before the computer age and complicated forest policy, I established my growth rate to be 1.9 per cent using longhand arithmetic. This has determined my annual cut ever since. Several years after I determined my growth rate researchers starting knocking on the door: a German forester came in at 2.0 per cent, an American at 2.1 and Forest Renewal BC at 2.1 with their computerized methods. So I was pretty close. In fact, I've been undercutting for all these years yet making a good living. In 53 years, I've taken out two-and-a-quarter times the original volume but still have 110 per cent of the original volume remaining. We carefully conserve the forest—our capital—and live off the interest. Now that's sustainable forestry. And it is a stark contrast to official forestry practice in BC which is specifically designed to liquidate the old-growth forest—our province's one-time only forest capital."

Certification is being used as a mechanism for promoting ecoforestry, with the Forest Stewardship Council being the preferred approach for the Ecoforestry Institute. During the last ten years, the Forest Stewardship Council has certified over 30 million hectares of forest and other 70 million hectares have been certified under other schemes. Although certification initially focused on large-scale industrial logging, community-owned forest enterprises and small holders, including indigenous people, are now being certified.

Ecosystem Based Management is another way in which ecoforestry is being characterized and this has been taken up on a large scale. The "Great Bear Rainforest" agreement announced by the provincial government of British Columbia in April 2001 between environmental organizations, First Nations communities, and forestry companies working on the north-central coast heralded an environmental coup, the reverberations of which will challenge the British Columbia forest industry. The agreement signalled the success of an international boycott campaign aimed at forestry companies singled out for environmentally degrading operations. As well, the agreement protects huge tracts of land from logging, including 96 458 hectares to preserve essential habitat of a rare white subspecies of black bear popularly known as the "Spirit Bear".

But the real environmental victory forged in the agreement lies in the announcement that all development activity in the plan area, which covers about 4.8 million hectares (equivalent to approximately 12 000 Stanley Parks) of marine, foreshore, and upland area on the north-central coast, will be based on the principles of Ecosystem Based Management.

Sustainable Forest Management

The broad introduction of the concept of Sustainable Forest Management can be traced to the so-called Forest Principles and Chapter 11 of Agenda 21, which were prominent outputs from UNCED. The guiding objective of the Forest Principles is to contribute to the management, conservation and sustainable development of all types of forests and to provide for their multiple and complementary functions and uses. Principle 2b specifically states that "Forest resources and forest lands should be sustainably managed to meet the social, economic, ecological, cultural and spiritual needs of present and future generations." It goes on to specify that: "These needs are for forest products and services, such as wood and wood products, water, food, fodder, medicine, fuel, shelter, employment, recreation, habitats for wildlife, landscape diversity, carbon sinks and reservoirs, and for other forest products." And that: "Appropriate measures should be taken to protect forests against harmful effects of pollution, including air-borne pollution, fires, pests and diseases, in order to maintain their full multiple value".

Although the Forest Principles form a "non-legally binding statement of principles", they bear the marks of a negotiated text with some repetitions and passages with ambiguous or very general text and are, in places, focusing on guidance for the establishment of an enabling framework for SFM, rather than specific guidance for application at the field level.

The concept of SFM has continued to evolve since 1992 through the international forest policy dialogue and a large number of country-led and ecoregional initiatives aimed at translating the concept into practice.

Nine ecoregional forestry initiatives or processes involving 149 countries, whose combined forest area equals 97.5 percent of the total forest area in the world, have been established since 1992 with the aim of translating the concept of sustainable forest management into practice. They continue to meet on a regular basis to further refine the concept through the development of criteria—or elements—defining SFM and sets of indicators for each of these aimed at facilitating monitoring of progress in the practical application of the concept.

Although evolving independently, these ecoregional processes are conceptually similar in objectives and overall approach and have shared information and experiences resulting in a convergence as regards the main elements constituting SFM.

The 2003 International Conference on the Contribution of Criteria and Indicators for Sustainable Forest Management (CICI) was held in Guatemala gathering representatives from all of the above processes, international organizations (including NGOs), government officials and experts in the field. Participants at this Conference agreed that SFM comprises the following seven common thematic areas:

- extent of forest resources
- biological diversity
- forest health and vitality
- productive functions of forest resources
- protective functions of forest resources
- socio economic functions
- legal, policy and institutional framework.

Of these, four are related to the environmental aspects of SFM and the remaining three to the social and economic aspects—the two other "legs" of SFM. As would be expected, the indicators vary widely among initiatives owing to differences in forest types and environmental, social, economic, political and cultural conditions. National-level criteria and indicators are being complemented by the development and implementation of criteria and indicators defined at the forest management unit level in a number of the processes as well as by other actors such as NGOs and the private sector.The degree of implementation of criteria and indicators at the national level varies considerably. In many cases, action is limited by the lack of trained personnel or institutional capacity for collecting and analysing information and for following up the development and implementation of improved management prescriptions based on the information obtained.

REFERENCES

Allen-Wardell, *et al.*, 1998. The Potential Consequences of Pollinator eclines on the Conservation of Biodiversity and Stability of Food Crop Yields. *Conservation Biology*, 12 (1).

Bugg, Robert L. and Picket, Charles H. (eds) 1998. *Enhancing Biological Control: Habitat Management to Promote Natural Enemies of Agricultural Pests.* University of California Press.

Edge, Daniel 2000. *Wildlife and Agricultural Ecosystems. Department of Fisheries and Wildlife,* Oregon State University.

European Preparatory Conference 2002, *Ecotourism in Mountain Areas: a Challenge to Sustainable Development.* Salzburg, Austria, 12-15 September 2001. International Year of Ecotourism and the International Year of Mountains.

Gliessman, Stephan R. 1999. *Agroecology: Ecological Processes in Sustainable Agriculture.* Ann Arbor Press 1999.

Imhoff, Daniell. 2003. *Farming With theWild: Enhancing Biodiversity on Farms and Ranches.* Island Press. Covelo, Ca. USA.

IUCN. 2002. Parks. Vol 12, no. 2, *Local Communities and Protected Areas.* Protected Area Programme of the World Conservation Union.

Mader, Paul *et al.* 2002. "Soil Fertility and Biodiversity in Organic Farming". *In: Science.* 296: 1694-1697.

13

Sustainable Agriculture in Small Islands

Small island developing States (SIDS) have limited resources and options for social and economic development. Consequently, careful attention should be paid to the manner in which resources are utilized so as to ensure that adverse cross-sectoral impacts are minimized and that inter-temporal benefits flowing from the exploitation of renewable resources are not jeopardized.

SIDS have in common a number of structural problems: their populations, and therefore their markets, are small; their resource base narrow, fragile and prone to disruption by natural disasters; they typically depend for foreign exchange on a small range of primary product exports; and they generally have limited local capital for productive investment. In short, their base for revenue generation is narrow. Agriculture has been the backbone of many SIDS economies, providing the main source of livelihood for population as well as being a major export earner.

In advocating policies and actions for island States, it should be recognized that while SIDS have a measures of commonality in terms of features and experiences, island States do not represent an homogenous group. This means that in terms of their individual situations and possibilities for social and economic development, not all island States, or groups of States, are equally advantaged or disadvantaged. Some SIDS, for example, have extensive marine fisheries resources on which to base national development planning while other States, which may be less well resource endowed can benefit from their proximity to markets in developped countries, not only as an avenue for some exports, but also as a source of tourism.

In general terms, the intensification and diversification of agriculture, forestry and fisheries production is constrained in island States by both natural and economic circumstances. These circumstances include, inter alia, a shortage of land, poor soils, a limited capacity of land to support intensive crop and animal production, difficulties with storage of products after harvest in tropical environments, infrequent and restricted capacity for moving products by sea and air to markets, and distance from, and access to, markets outside the region.

There is now international recognition that island States face particular problems with respect to primary sector development and management. This recognition is increasingly reflected in recent international instruments of all types. Among instruments that have a direct bearing on food security and in which the situation of island States are noted specifically are the 1995 Kyoto Declaration and Plan of Action on the Sustainable Contribution of Fisheries to Food Security are the 1996 Rome Declaration on World Food Security, and the World Food Summit Plan of Action.

Paragraph 28 of the Basis for Action of the World Food Summit Plan of Action notes that "Small Island Developing States face the threat of land loss and erosion due to climate change and sea level rise and have particular needs for their overall sustainable development. Improvements in transportation, communication, human resources, stabilization of income and higher export earnings will increase food security in these countries. ..." The situation of island States is therefore special in terms of broad initiatives to be adopted to promote long-term, sustainable food production in support of food security.

The following strengths, weaknesses, opportunities and threats that confront SIDS have been identified:

Strengths:

- extensive traditional knowledge of the environment and natural resources and traditional management practice of resources;
- possibilities for permanent outward migration and temporary migration for work as a means of reducing pressure on resources and available employment opportunities;
- important cash remittances from abroad from islanders who have migrated overseas or who have found work outside their countries;
- large EEZs in comparison to the size of land area and populations; and
- established regional and international cooperation as is evident through general and specialised regional groupings and AOSIS.

Weaknesses:

- limitations associated with smallness of scale;
- difficulties in maintaining and improving the national contribution to food security;

- a lack of national capacity to develop and implement policies for the sustainable use of agriculture and natural resources which is in part exacerbated by outward migration by skilled islanders;
- limited national infrastructure to support increased and diversified production from primary sector activities;
- geographic location and resulting disadvantages for some island groupings which affects their capacity to develop and market new products and to benefit generally from market opportunities available. Relative isolation for some islands also inhibits the relative availability of energy, and transport and communication facilities. This isolation and lack of facilities can make these islands unattractive as tourist destinations; and
- in forestry, a long lead-time is required to bring forests into production.

Opportunities:

- SIDS present opportunities for increasing primary sector production and diversification. These opportunities may be expected to make economies more buoyant and robust, to provide additional employment opportunities and possibly to generate additional foreign exchange;
- mineral development, which although can be problematic in SIDS from the environmental perspective, can generate important revenue to support national development;
- dietary improvement as a result of increased production and diversification;
- economic diversification and development and improved income earning possibilities;
- tourism development and strengthening, including the possibility of promoting small-scale eco-tourism;
- enhanced policy coordination and integrated management so that links and impacts between agriculture, forestry, fisheries and other primary and secondary activities are assessed and managed more effectively;
- strengthened regional cooperation among island States as well as opportunities for inter-regional cooperation and exchanges;
- measures to improve value-added processing and related employment and national income benefits; and

– privatisation of existing investments in the primary sector and the promotion of private sector investment.

Threats:

– fragile environments that may take many years to restore themselves following devastation by hurricanes etc;

– generally, high rates of population increase, especially in the South Pacific and some countries in the Indian Ocean. This increase in population exerts pressure on all resources in island States and has major implications for the sustainable management of natural resources and food security;

– impacts of climate change including sea level rise and the potential loss of land and infrastructure which is primarily located in the coastal zone and reef bleaching and habitat loss for reef associated species of fish and abnormal migration patterns for offshore pelagic species. In the extreme, the existence of some low-lying atoll States may be threatened completely by sea-level rise;

– vulnerability of island States to natural hazards such as cyclone, volcanic eruptions and earthquakes which causes loss of life, major economic losses and disruption, including damage to, and loss of, infrastructure and impacts on crops, and other resources. Moreover, because of the size of most SIDS and the fact that many island States lie within natural disaster belts, such natural calamities normally affect the entire country when they occur and even neighbouring countries in some instances;

– effects of forest fires and the introduction of pests of exotic species;

– reduction of biological diversity through the loss of endemic species. Risks of extinction of some species are high in SIDS because of the their small size which prevents the setting aside of areas for strict protection purposes;

– political and economic instability in some island States; and

– intense competition, especially near urban centres, for land and water resources and space in the coastal zone.

TRENDS IN AGRICULTURAL SECTOR

SIDS are looking for opportunities to diversify their economies and especially the agricultural sector in order to move toward more marketable crops, to increase foreign exchange earnings, to maintain their significant agricultural basis, to increase their degree of food security and self-reliance by exploiting their resources more rationally and sustainably and to prevent unemployment from worsening. Diversification is one means of improving resource use, though implementing diversification policies which support return to home-produced foods to substitute some imports, cultivation of export crops for niche markets, changes in farming systems, alternative uses of forest products will be a slow process.

The small size of many SIDS makes competition for land severe. Large commercial plantations were established under the colonial system, and continue today in the more fertile plains, forcing many small farmers to cultivate the poorer soils in hilly regions. This, combined with poor land use practices, has resulted in deforestation and erosion. In low-lying coastal areas, land well suited for agriculture is under severe pressure from expanding housing, industry, and tourism. Undefined land tenure in a number of SIDS renders farmers unwilling to invest in agricultural development.

Generally, water resources are scarce, often limited to thin sheets of freshwater floating on seawater, recharged by rainfall. Poor management of watershed catchment areas and of water flowing through agro-ecosystems, particularly irrigation supplies, can lead to water scarcity and pollution of downstream supplies. Water resources for irrigation purposes have particularly been developed in the Caribbean. Population pressure, expanding tourism, and reduced precipitation of recent decades, have resulted in the over-exploitation of aquifer resources leading to subsequent salt water intrusion in lands near the shore and salinisation of groundwater, as well as less land and water being available for agricultural purposes.

The most important food crops grown in SIDS are starchy staples such as cassava, sweet potato, yam, potato, cocoyam and taro and plantains and bananas which are mostly produced by small-holders. Grain crops (e.g. maize, wheat, rice etc.), though consumed widely, are only grown on a limited scale. Due to the rapid rate of urbanization and the relatively low level of consumption of root crops and musa species in towns, these staple crops were increasingly substituted by imported cereals especially wheat and rice in the last decades. However, root and tuber crops remain the main source of nutrition in the rural areas of many SIDS. The major export crops of SIDS include banana, sugar cane, cocoa, coffee and coconuts. In many SIDS these crops are the most important net foreign exchange earners. Agricultural exports represent between 20 to 90% of the total trade exports. Livestock in SIDS is mainly based on poultry and pigs raised under traditional small-holder conditions, whereas commercial cattle production is a more recent development.

Issues and Constraints

SIDS are dependent to a great degree on the international trading system. Many of these countries receive some form of preferential access to developed-country markets. The EU and the US, for example, grant preferences to a number of SIDS under the Generalized System of Preferences. The EU provides additional preferences to 26 SIDS under the

Lome convention for ACP countries. This dependence means that the achievement of sustainable agriculture is more complex than for countries which have to rely less on the world market. The integration of SIDS into the global economy is particularly difficult due to the following constraints:

- limited population size making economies of scale and labor specialization impossible;
- emigration of skilled personnel, and a high degree of dependence on foreign capital and external assistance;
- a small monetized sector with local private investments limited to low-risk activities such as trade and services;
- a heavy burden on the public sector faced with limited domestic income-generating capacity reliant on sources such as import duties, fishing rights and remittances from nationals working abroad;
- a narrow resource base creating a dependence on imports of consumer and capital goods;
- export limited to primary products from agriculture, forestry, fisheries and mining prone to considerable fluctuations in output and price.

The sector in many SIDS generally comprises:

- a large number of traditional small-holder farming systems that practice mixed cropping (root crops, pulse, vegetables, and perennials) mostly for home-consumption with a view of spreading their risks. They have limited access to traditional credit sources;
- a small number of small commercial farms of high capital base and easy access to traditional credit sources;
- a small number of large farms with high resource base and access to traditional credit, these dominate the agricultural sector and account for a large portion of production of traditional export crops (banana, sugar, coconuts, cocoa, coffee etc.);
- a small number of large farms that are currently idle. This group is beginning to consider investment in traditional commodities.

Production Constraints

Farmers have traditionally used few purchased inputs for their low-input/low-output production system. The introduction of new commercial crops (cocoa, bananas etc.) and traditional mono-cropped root crops has increased the demand for fertilisers, insecticides, improved tools and credit, while the rising cost of labour has led to the use of herbicides for weed control. Often these inputs are costly and not readily available.

The availability of farm labour is a persistent constraint. This is due to the emigration of a large part of the young male population in many SIDS and the alternative higher-income employment opportunities in commercial farming, urban areas and the tourist sector. As a consequence the percentage of female-headed households has increased. The cost of labour has increased and acts as a constraint on agricultural development. The high level of remittances from emigrants is a key factor in driving up local labour costs.

Planting material availability is generally inadequate to meet demand particularly if rapid expansion of production is to be undertaken. There is a strong dependence of SIDS on imported seeds for many crops. There is also a fundamental problem related to quality control due to the continuing and extensive use of planting material from preceding crops rather than certified and clean source of material.

With respect to plant protection SIDS have to cope with the stringent quality demands of agricultural produce by importers in industrialised countries or with the changing requirements of local food consumption especially in the tourist industry. This and the favourable climatic conditions for pest and diseases and the high cost of labour have led to the use of pesticides as a standard practice in the agricultural production systems. However, there is a lack of an efficient pest and pesticide control and monitoring program. Only few SIDS are participating in the Convention on the Prior Informed Consent (PIC) procedure which contains provisions for the exchange of information between exporting and importing countries about potentially hazardous chemicals.

Improper selection of produce for harvest and improper post-harvest handling, packaging and storage techniques due to the lack of knowledge and appropriate technologies, contribute to the substantial post-harvest losses in many SIDS. Exports of smallholder-grown root crops, fruits and vegetables are constrained by frequent failure to meet the quality and quarantine standards of importing countries and the irregular supply.

SIDS present the livestock producer with a series of disadvantages but relatively few advantages. Apart from keeping a few household pigs and poultry, more commercially orientated livestock keeping is relatively new. Geographic isolation and associated high transportation costs seriously affect the financial viability of any livestock enterprise dependant on imported inputs e.g. broilers and intensive egg and milk production. Unless there is a niche market, such as tourism, willing to pay a premium for a fresh product these enterprises are unlikely to be financially viable.

Major constraints include poor genetic stock, inadequate or poor nutrition, poor animal health services, high costs of purchased feed, problems in procurement of commercial feed, inadequate management of breeders and hatcheries, and labour shortage. Due to these major constraints, the scope for developing livestock production appears to be limited. There are possibilities of intensification in poultry in some of the larger islands. Greater support is needed to develop the livestock feed industry and to identify and establish market structures, as well as support services for smallholders.

Market Constraints

There are traditionally two principal markets:

— the domestic market for food and other products of the production system, and

— the world market for tradable commodities (e.g. cocoa, coffee, copra, banana etc.).

The private enterprises and to a lesser extent, the semi-government organisations have developed relatively efficient marketing systems for the commercial crop sub-sector. The domestic marketing system which is usually based on non-institutional channels, mainly private intermediaries, is generally poorly developed. Due to the small and static population, remoteness of farms, poor infrastructure and high transport costs, low investment in market research, and poorly developed market information systems efficient marketing systems are difficult to establish.

There is a limited knowledge of the market place by producers and exporters. This includes the organisation and structure of the markets dealt with as well as the nature of the operations and the participants involved in the markets. Information on competitors (equipment they use, sales force, prices, products, terms they offer etc) is generally lacking and makes an assessment of the competition difficult. Marketing economies of scale or the gains made by larger entrepreneurs are not available to the small volume producers and exporters in the SIDS. In reviewing the current practices, the habit of scattered uneconomical small-scale production of a wide variety of crops is evident. The provision of better market information is needed for more viable long-term production decisions.

A key constraint to be addressed in promoting exports is the high cost and limited availability of air and sea transportation systems, the lack of or poor fresh produce handling facilities, and inadequate or inappropriate transportation services for non-traditional produce. Plant quarantine issues also need to be addressed as they restrict market-access for produce.

Institutional Constraints

The Ministries of Agriculture have the important role of implementing governments' agricultural policies and programs at the national level. The research and extension services are key to increase the productivity and output of agricultural enterprises.

Technological progress is markedly different among crops. Predominant traditional food crops such as cassava, taro and sweet potatoes, developed over the centuries, are relatively efficient in both production of calories and financial returns to labor input. Yields of coconuts and introduced crops such as cocoa, coffee and bananas have remained low despite the fact that most research and extension efforts in the past were directed at such introduced commodities. Part of the difficulty in successfully establishing introduced crops in the smallholder sector is because of lack of knowledge about how such exotic commodities fit into the traditional farming system.

In most SIDS agricultural research, mainly government-funded, is constrained by very limited human and financial resources. It is weakly linked to extension, the farmer, and other sectors such as agro-processing and tourism. Few SIDS have clearly identified research priorities, which are relevant to the smallholder sector. There is a considerable need to train local manpower in research and technology development at post-graduate level.

Marketing Boards: Many SIDS have marketing boards, which are statutory marketing authorities. They are geared primarily towards generating income from activities related to the export of traditional commercially-produced agricultural commodities and not generally available for use by the farmers and private exporters of these or other non-traditional agricultural products. They have not yet responded competitively to the growing and more efficient, private sector exporters of agricultural produce.

Farmers Organizations: The impact of farmers organizations has been limited because they generally suffer from weak management, low level and quality of services, low participation of members and inadequate financial resources. However, there are some farmers organizations (e.g. in the Caribbean) where private exporters are the backbone of the export trade in non-traditional crops. These small and medium size operators tend to avoid formal credit institutions because of fear of debt and the associated collateral requirements. They prefer to seek informal credit services. In order to further commercialise agricultural operations there is therefore a need to change negative attitude to credit and a need for the credit institutions to devise financing schemes for small/medium commercial operators.

Diversification

The current agriculture in SIDS is characterised by a combination of large-scale commercial production of cash crops and an important smallholder sector which produces crops primarily for local consumption. Agriculture in SIDS is in transition, driven by changing world markets, trade imbalances, the quest for food security and growing human population. Given the potential changes in trade preferences for traditional export crop industries (e.g. banana) and the objective of food security, agricultural diversification and intensification can contribute strongly to broaden the potential source of income of farmers. Exports of some basic raw food commodities, certain high-value fresh tropical fruits, niche-market fruits and vegetables during the northern winter season, and value-added processed foods are extremely important to small-, medium- and large-scale producers and processors and to the overall economy. Several SIDS are well suited to fresh fruit and vegetable production and some of them are relatively close to markets in Australia, New Zealand, Japan, USA and Europe. There is also a potential for inter-country trade or import substitution. A more diversified production structure may provide greater stability in export earnings and promote import substitution. Provided a range of appropriate policies is implemented, diversification can contribute to sustainable agriculture, the sound utilisation of natural resources and the protection of the environment.

SIDS have some opportunities for increasing agricultural production and growth, and farmer's income could be improved, provided policies support advantages and address constraints. A precondition for tackling the constraints is to ensure that land, labour and capital markets are operating efficiently. The key characteristics of successful activities which could be developed in the future are:

- returns to farm labour must be above average wage rates; labour intensive activities must be avoided unless returns are high;
- exports will be as fresh produce (e.g. off-season supply), or with processing that can be done on the farm, is not expensive in relation to product returns, or can be done centrally, without the need for substantial economies of scale (e.g. specialist products);
- produce must be storable without quality loss within the normal limits of transport;
- produce must be high value in relation to transport cost;
- quality must be according to consumer requirements, including presentation and packaging and quarantine.

A successful strategy for agricultural diversification and intensification should comprise the following major activities:

LAND RESOURCE MANAGEMENT

Sustainable development in SIDS is directly dependent on or influenced by, the planning and utilisation of land resources. Mechanisms for an integrated planning and management of land resources including inland and coastal areas are of great urgency and importance for SIDS. FAO is promoting an improved approach to land resource management which emphasises the integration of physical, socio-economic and institutional aspects of land use, and stresses the need for active participation of all stakeholders in decision-making. It has also been involved in the development of national computerised land information systems including a comprehensive Geographic Information Systems database in the Caribbean. With this specific expertise, FAO could assist decision-makers and land users in increasing the effectiveness of land resource planning and management at the national and local level.

Sustainable Farming Systems

In order to protect the fragile ecosystem of SIDS sustainable production systems need to be developed. The following actions can be proposed:

- Traditional or moderate input systems. Develop sustainable agro-forestry systems to: raise and diversify production, improve soil fertility, prevent soil loss and environmental degradation, and reduce dependence on external inputs;
- Intensive high-input agricultural systems on lowlands. Introduce short- duration cover crops and legumes to improve soil fertility and structure, conserve moisture, reduce build-up of weeds and pests, reduce reliance on imported chemicals and fertiliser, minimise environmental degradation, and increase green fodder availability;
- Farming systems research. Appraise socio-economic issues and feed this information into cropping trials and extend technology to the farming community using a farmer-to-farmer approach. Research should also focus on tree crops, new crops to be introduced, mixed perennial cropping systems, the multi-crop smallholders that utilise agroforestry systems, and livestock raising which optimises production and maintains soil productivity. In the future, a more holistic approach is needed with interdisciplinary and usually multi-institutional studies of ecosystem management, biological interactions of mixed crop, tree and animal production systems, including aquaculture. Assessment of indigenous knowledge and traditional production systems, and development of appropriate technologies for high and/or low potential areas needs to be carried out.

The bottom line for adoption of improved technologies is income generation, the economic merit must guide choices.

Seeds and Plant Genetic Resources

Intensified and diversified crop production requires a strong seed program either at regional or sub-regional level. Elements of such strategy include:

- the development of regional technical capability for seed supply to enhance, in a selective manner based on the country situation, the establishment of seed programmes;

- the development of appropriate seed policy to enhance national and regional efforts in plant genetic resources utilisation and seed supply; and
- development of a germplasm information network to link up the separate islands and bridge information gaps in the region.

Irrigation

Exports of traditional crops have declined in recent years and productivity of non-traditional crops is low. In order to combat the serious decline in revenue, irrigation is an opportunity to reduce the unit cost of growing a crop by increasing yields and improving quality, and diversifying into other crops. In islands where water is scarce, integrated water management systems can be promoted. Localised irrigation systems, either high tech or low-tech can be introduced for high-value crops. In the highlands of arid or semi-arid islands, various types of water harvesting and runoff farming techniques can be introduced, mostly for human and animal consumption, and to a lesser extent for gardening.

Plant Protection

Current pest control strategies based mainly on the use of pesticides should be reviewed. The practice of Integrated Pest Management (IPM) is an alternative strategy for effective, efficient, balanced and environmentally safe pest control. SIDS are by definition relatively isolated areas of limited size were biological control can be effective, relatively easy to implement, because pests and diseases cannot spread easily to neighbouring territories, and their impacts on the natural ecosystem are often more predictable. Production could be undertaken in official Pest Free Areas (PFA) so as to meet special needs of niche export markets. Pest eradication programmes should be seriously considered when dealing with important crops and a small number of pests with a simple life cycle and infection/infestation process. Other areas of intervention are:

- establishment of harmonized pesticide legislation and registration requirements to establish a common legal framework for the importation and distribution of pesticides;
- monitoring of pesticide residues in the environment and in agricultural produce;
- active participation and joint decision making within the PIC procedure; and
- verification of the existence of obsolete pesticides and their disposal.

New Crops

SIDS in general have a strong comparative advantage in the production of some tropical fruits (e.g. papaya, plantain, mango, pineapple, watermelon, dasheen, tania etc.), tuber/ root crops (e.g. taro, yams, sweet potato, cassava), nuts and spices (e.g. canarium nut, Brazil nut, terminalia nut, vanilla, black pepper) vegetables and cut flowers. The potential for diversification into and intensification of these new crops (possibly under organic farming which excludes the use of agro-chemicals) for niche markets should be explored. Increased production would be absorbed both in the domestic market (e.g. local tourist market) either to meet increased demand or substitute for imports, and in export markets. Both traditional and non-traditional (new) crops should be cultivated on the basis of Integrated Crop Management and Conservation Farming principles in order to achieve sustainable and environmentally friendly production.

Post-Harvest Handling and Agro-Processing

Since a large proportion of agricultural produce such as root and tuber crops, fruits, and vegetables is locally marketed fresh, development of appropriate post-harvest processing and storage technologies are extremely important. Farmers, extension staff, middlemen and retailers in the local markets generally lack information on post-harvesting handling of fresh produce, proper packing, transportation and storage of the produce before sale. There is therefore a need to promote simple and low-cost technologies, preferably using local resources, for handling and packing fresh produce, to reduce damage during transportation and storage.

Considerable opportunities exist for expansion and diversification of agro-processing industries to produce added-value products to generate employment, which is much needed. Appropriate technology and expertise in agro-processing could be developed with a particular focus on developing small-scale, low-cost agro-processing ventures to increase the value and market potential of products (e. g. root crops, banana, breadfruit, tropical fruits, spices and coconut).

Animal Production and Health

Production and productivity of small-scale poultry and pig production can be increased using locally-available feeds. Development of feed mills, based on local ingredients could encourage semi-intensive indigenous livestock industry. Diversification, through the introduction of alternate livestock, such as ducks and pigeons, may be a possible option. The integration of livestock into mixed-farming and tree-crop systems has the potential to optimise use of crop residues and vegetation on uncultivated land, as well as to assist the recycling of soil fertility. Management skills, feeds and pastures in order to realise the full potential of improved breeds need to be upgraded. Niche markets for cheese and yoghurt as a more versatile product than fresh milk could be developed. Animal health services should be supported, local staff trained, and specific disease eradication programmes (e.g. Tropical Bont Tick in the Caribbean) be promoted.

Marketing and Credit

The marketing strategy for SIDS agricultural exports to hard currency areas will hinge on joint marketing, promotion of commodities being sold in niche markets, monitoring the need for change in operating strategy as commodities enter into mainstream markets and the building of strategic alliances with market participants (producer/exporter/ importer alliances). Joint marketing services with strong private sector participation, should be knowledgeable of the markets, set standards, provide information and co-ordinate decisions on buying raw materials, transportation, packaging, training, quality, etc. in order to be more competitive. Existing Marketing Boards and private exporters could be seen as the agencies best placed to provide:

- market intelligence and information;
- exporter facilitation, and
- develop farmer organisations (FOs).

The development and strengthening of FOs is an important aspect of ensuring the sustainability of the production of non-traditional agricultural commodities. Selected FOs should have commercial orientation, operate economic size units and secure land tenure. Exploiting organic agriculture markets may be possible in some cases.

Availability and accessibility of credit will be critical to achieving diversification and intensification of non-traditional commodities. Lending conditions should be sufficiently broad and flexible to allow for the kind of investments associated with on-farm development works envisaged including those for fruit tree crops with relatively long gestation periods.

Institutional Development

The diversification and intensification process requires modifications to the institutions (public and private) engaged in the process. The organisational structures have to ensure that there are formal linkages at all levels of the diversification effort in order to establish the framework for a high level of co-operation and co-ordination of the national (and possibly regional) diversification policies, plans and programmes, and that there is a strong commitment to the diversification thrust which will result in a focused, co-ordinated approach involving all the key individuals, institutions and agencies.

At the national level, the Ministry of Agriculture in particular has a critical role to play in creating the environment for and facilitating the production of the non-traditional crops which fall under the diversification plan. It is responsible for:

- providing support for the production of non-traditional crops;
- mobilising the resources to ensure success in production;

- advising farmers on crop types and production systems;
- educating farmers to undertake farming as a business;
- providing technical services in support of farm production operations;
- motivating farmers to produce non-traditional crops to meet market demand.

The current functional organisation of the ministries in many SIDS does not lend itself to providing the focused, crop-specific approach required to drive the agricultural diversification thrust. Agricultural diversification requires a pro-active, project-oriented and multi-disciplinary approach to develop the production/marketing systems required for the targeted non-traditional agricultural crops.

Human Resource Development

Skills of all actors (e.g. farmers, exporters, administrators and extension personnel) need to be permanently upgraded through specialised training and capacity building in the field of agronomic, financial management and marketing expertise. Particularly extension staff requires a reorientation to commercial farm management with attention to be paid both to productivity improvement and better business planning and to related training of farmers in improved organisational and managerial skills. Training and institution building will often be more effective if they are undertaken on a regional basis.

Technical Co-operation

Exchanging experience and information among developing countries and taking advantage of complementarities to upgrade skills through Technical Co-operation among Developing Countries (TCDC) is highly cost-effective. TCDC has great potential that developing countries have not yet fully exploited. Regional groupings and mechanisms could promote and implement TCDC. Inter-regional technical co-operation among SIDS focusing on e.g. training, technical exchange, and attendance at regional meetings could be strengthened and FAO could play a supporting role.

Information Systems

Data bases, essentially for planning purposes, are generally weak in SIDS. Although the methodology for collecting, processing and analysing data for food and agriculture is available in most SIDS they lack adequate statistical organisation and personnel, and understanding and co-ordination between statistical offices and economic analysis, planning and decision-making agencies. Particular attention should therefore be paid to establishing an institutional interdisciplinary framework. Additional data, new methods of data analysis, and appropriate tools to integrate environmental, social and economic considerations in decision-making are required.

FISHERIES AND AQUACULTURE

Regional differences between SIDS, and even between SIDS of the same region, affect natural resource endowments for inshore, offshore and inland capture fisheries as well as the capacity of island States to promote the development of aquaculture. It is therefore difficult to make generalisations about fisheries and aquaculture in SIDS, and any such generalisations must be premised on caution.

Fish are a critical source of animal protein for populations in island States. Fish therefore figures heavily in the food security equation in these countries. Largely by necessity per caput fish consumption rates in SIDS are high by international standards. In many island States women play a key role in promoting food security through their activities in the fisheries sector. While women do not normally engage in fishing per se, except from shore (i.e. they do not fish from water bound craft), their activities within the sector include the processing of part of the catch (though generally there are not strong processing traditions in SIDS), and marketing. This activity extends to the distribution and sale of both inshore catches from artisanal and small-scale fishermen, and the by-catch of industrial fleets (e.g. trawl 'trash' fish and discardable species from tuna fleets). This industrial catch can be important for food security in urban areas in island States as the by-catch species are usually available to the most disadvantaged groups at an affordable price (e.g. Solomon Islands).

In artisanal and small-scale fisheries in SIDS the incidence of discarding of catch is low, with catch utilisation reaching virtually 100 percent. However, in industrial fisheries, and in particular in those fisheries that are not based in island States, the incidence of discarding of by-catch is, by necessity, higher than for the other locally-based fisheries.

The Kyoto Declaration and Plan of Action on the Sustainable Contribution of Fisheries to Food security makes specific reference to SIDS. In this regard the Declaration and Plan of Action note that many developing countries, in particular low-income food deficit countries and SIDS (i) face major challenges in ensuring the sustainable contribution of fisheries to food security; (ii) require technical and financial assistance in order to enable them to realize the sustainable contribution of fisheries to food security and social and economic development; and (iii) should cooperate among themselves in order to achieve the contribution of fisheries to food security.

In general, inshore capture fisheries adjacent to centres of urban population are heavily fished if not overfished, while inshore fisheries in outer islands tend to be only lightly exploited because of lower population densities and the paucity of market infrastructure to move product to urban markets. Species harvested in coastal areas are usually mixed, with some species (e.g. deepwater snappers) being of high quality and value. A range of traditional and introduced harvesting techniques are used in these fisheries, with motorised craft being common where there is access to fuel and basic engine maintenance. The use of destructive fishing practices (e.g. dynamite, cyanide and bleaching), although outlawed in most SIDS, continues to be common in some islands.

A range of environmental considerations including the effects of El Niño which can lead to abnormal fish stocks behaviour for offshore resources and coral reef bleaching and loss of habitat for reef associated species, adversely impacts the capacity of marine capture fisheries to support productive and sustainable fisheries. These considerations are exacerbated by overfishing and the need for realistic and socially acceptable conservation and management arrangements. It has been shown in many developing countries (including SIDS) with increasing regularity, that conventional approaches to fisheries management have not been effective in promoting sustainable resource utilisation. As a result, innovative approaches to management that build on traditional institutions and conservation and management knowledge and practice are now considered to be more appropriate. This recognition has given rise to the promotion of community-based fishery management, which has been successfully demonstrated to be implementable in those island States populated by indigenous communities where marine use tenure systems remain intact, or largely intact.

Offshore capture fisheries and associated processing provide major economic benefits for the island States of the South Pacific and the Indian Ocean, and to a lesser extent in the Caribbean, although there is increasing interest in such fisheries among several Caribbean States. Tuna and associated species are the main stocks targeted, usually by foreign fleets from Asia, Europe and the United States of America. In the South Pacific about 1 million tonnes of tuna is harvested annually. This catch has a 'wet', on board value of around US $1.6 billion.

In this region some 80 percent of the catch is taken within the exclusive economic zones (EEZs) of SIDS. In the Indian Ocean the tuna and other pelagic catch is about 1.1 million tonnes per annum with a value of some US $ 2.3 billion. An estimated 50 percent of this catch is taken on the high seas in this region mostly by non-Indian Ocean States.

In both, the South Pacific and the Indian Ocean only a small proportion of total tuna catches are taken by SIDS' fleets. In the Caribbean only a small proportion of pelagic resources are taken within the EEZs of island States.

An important policy consideration for all SIDS where there is significant offshore fishing, either within areas of national jurisdiction or on the high seas, is the need to secure additional benefits from the exploitation of regional fish stocks. Some island States (e.g. Seychelles, Federated States of Micronesia, Palau and Papua New Guinea) have taken progressive measures to increase their economic linkages with offshore fisheries by, inter alia, encouraging the development of national fleets, either through direct investment in vessels or through flag changes. In addition, steps have also been taken by some island States to facilitate the use of their ports, and where necessary, to make the generation of secondary economic benefits (e.g. the transhipment of fish in port rather than at sea) conditional on fishing access to their EEZs. These measures, in many instances, have turned out to be mutually beneficial for both the island States and the foreign fleets.

Both inshore and offshore marine capture fisheries are central to tourism development in SIDS. Inshore resources are important for water sports, while the sports fishing industry hinges on the availability of offshore resources. To retain their popularity as water sport/sport fishing destinations, SIDS must ensure that policies are in place to carefully and sustainably manage tourism: on the one hand there is a need to facilitate the maintenance of fish stock abundance, and on the other, a requirement to prevent tourists from engaging in such practices as habitat and environmental degradation (e.g. removing corals, shells and other animals from reefs).

Inland capture fisheries are not of major importance for most island States. However, in some of the larger island countries (e.g. Fiji, Jamaica and Papua New Guinea) catches of finfish, freshwater prawns and shellfish contribute in a significant way to food security and provide economic opportunities for self-employment, often for women. However, the prospects for increasing production from inland fisheries is SIDS is not great, though like their marine counterparts, inland capture fisheries are in need of management.

Regional cooperation in marine capture fisheries among SIDS is well developed in all regions. Island countries recognize that given their limited national financial and human resources, cooperation is essential in fisheries given that many of the stocks utilised are shared and fisheries challenges are faced. Cooperation is pursued through regional organizations such as the Pacific Community and Forum Fisheries Agency in the South Pacific; the Caribbean Community (CARICOM) and the Organization for Eastern Caribbean States (OECS) and the Western Central Atlantic Fishery Commission (WECAFC) in the Caribbean; and the Committee for the Development and Management of Fisheries in the Southwest Indian Ocean of the Indian Ocean Fishery Commission (SWIO/IOFC).

Regional fisheries cooperation ranges from the sharing of fisheries and related information, joint training and research programmes, to collaboration in monitoring, control and surveillance (MCS). Indeed, regional cooperation in the area of MCS, as a means of responsibly strengthening fisheries management, is being pursued as a matter of importance in both the Caribbean and South Pacific regions. Efforts to strengthen regional fisheries cooperation at all levels and in all areas should be actively encouraged among SIDS grouping given the benefits that such cooperation can bring.

Aquaculture is being promoted in SIDS to augment fish production (e.g. tilapia culture in Jamaica and Fiji for food and for export) together with other marine products such as pearls (e.g. Cook Islands), seaweed (e.g. Kiribati and Tonga) and aquarium fish. Reef enhancement is being tested with a variety of organisms in various states, but its viability is still to be demonstrated. The culture of these products provides national employment opportunities and revenue where exports are involved. However, the extent to which island States can participate in culture-based fisheries, and the nature of the products that are economically viable for culturing, is determined largely by their proximity to international markets and the availability of transport services. Caribbean States are generally better placed than SIDS in other regions to engage in culture-based activities for export in view of their adjacency to large North American markets.

In aquaculture a number of regional programmes and networks exist to facilitate information flows and dissemination of technical information. Notable among these regional aquaculture initiatives are the FAO South Pacific Aquaculture Development Programme, the International Center for Living Aquatic Resources Management (ICLARM), the Australian Centre for International Agricultural Research (ACIAR), the Japan Overseas Fisheries Foundation (OFCF) and the Japan International Cooperation Agency (JICA) in the South Pacific. In the Caribbean a number of institutions have been active in promoting regional activities in aquaculture, including CARICOM, the US National Marine Fisheries Service (NMFS), and the US Agency for International Development (USAID).

Fisheries Sector Constraints

A number of factors can be identified that constrain the development and manage of the fisheries sector in island States. These factors include, inter alia:

- A lack of institutional strength and capacity on the part of the public sector to manage the fisheries sector. Improving capacity is seen as being a vital pre-requisite for enhancing the overall management of the sector and thereby realising its full contribution to national food security and social and economic development. It is being increasingly recognized that as a result of high staff turnovers in small national fisheries administrations, institutional strengthening and capacity building, particularly in terms of human resource development, requires an on-going and long-term commitment on the part of island governments and technical assistance agencies.
- Difficulties associated with improving inshore fisheries management, using traditional knowledge and practice through community management and relocating fishing effort to offshore areas in those cases where pressure on resources is less intense, must be overcome if sustainable fisheries are to be achieved. To address constraints of this nature a coordinated set of policies is required that would involve, for example, training, extension, new vessel design, finance, and even new port

facilities. As part of improving fisheries management, islands States, individually and collectively, are urged to take steps to implement the Code of Conduct for Responsible Fisheries, and to ratify and accept respectively, the UN Fish Stocks Agreement and the Compliance Agreement, and to ensure that action is taken to give effect to these Agreements.

- High post-harvest losses due to poor fish handling and limited and underdeveloped national and international fish marketing arrangements constrain fisheries utilization. For inshore catches the main marketing problem is moving fish from outer areas to urban centres, while maintaining quality and being able to offer fish at a reasonable price. This problem is not unique to SIDS but there are particular difficulties that may not be encountered, or which are less pronounced, in continental States.
- Inadequate safety regulations and systems for fishers, especially those that operate beyond the immediate inshore areas inhibit a better distribution of fishing effort. Many islanders are lost at sea each year because fishing in bad weather, fishing with inappropriate craft and gear and a general lack of knowledge about safety at sea (e.g. no or insufficient life saving equipment is carried on fishing craft). The prospect of loss of life at sea constrains fishers from extending their operations seawards and away from heavily exploited inshore fisheries.
- Immature national fishing industries, many of which are public sector owned, and a weak private sector affects the extent to which the fisheries sector can contribute to national development. SIDS would benefit from assistance to privatize publicly owned investments and to put policies in place that would facilitate a strong and vibrant private sector. While recognizing that many SIDS have low absorptive capacity for investment because of their size and indeed may have difficulties in attracting investment for that reason, well defined policies to attract national and foreign investment, supplemented by legislation that provides reasonable protection for investors, should be vigorously encouraged.
- The current lack of aquaculture and inland fisheries development in small island States constrains the overall contribution of the fisheries sector to food security and national economic development. Most SIDS have yet to realize their full potential for the development of aquaculture and inland fisheries. In particular, measures to promote sustainable aquaculture and reef enhancement are required, while ensuring that inland capture fisheries are harvested in a rational manner.

Prospects for Increasing Production

The prospects for increasing production from marine capture fisheries resources in island States vary regionally:

- In the Caribbean region many of inshore resources are already heavily fished, if not overexploited, so the chance of substantially increasing coastal production is not promising. In this region production increases are likely to come from offshore resources (e.g. pelagic resources), but these resources, which are subject to management by the International Commission for the Conservation of Atlantic Tunas (ICCAT), are targeted by foreign fleets.
- In the Indian Ocean inshore fisheries are also heavily fished, except for countries such as the Maldives, where reef related stocks are viewed as being important for tourism. In other countries improved marketing of inshore product could stimulate production within sustainable bounds. However, any significant increase in production would have to come from offshore fisheries, which are also being exploited by foreign fleets, and subject to management by the Indian Ocean Tuna Commission (IOTC).
- Inshore fisheries in the South Pacific are also generally overfished except in more remote areas and outer islands. There are prospects for increasing production from these fisheries, but there are major difficulties in marketing. The greatest potential for increasing production comes from offshore pelagic resources, many of which are considered, at the present time, to be underexploited.

The fisheries sector in many island States does not offer significant prospects for diversification in terms of production. However, an exception is the growth industry in the South Pacific whereby locally caught fish (e.g. snappers and tunas) are being air-freighted to markets both in and outside the region. Snappers, for example, are being shipped to markets in Australia, New Zealand and United States of America, while high-grade tuna is being airlifted to sashimi markets in Japan and the United States of America.

With respect to the diversification of value-added processing, some possibilities exist and new markets are opening up. A study on this subject recently commissioned in the South Pacific has identified a wide range of new products, largely based on tuna, appropriate for processing in island States. Likewise in the Caribbean, WECAFC is involved in promoting new fish products for human consumption, partly due to the rising demand for fish and fishery products in that region.

Planning and Management

Considering the small geographical size of many SIDS, externalities of any economic activity will most likely be felt on the whole island, especially in coastal areas. Increasing levels of economic activities further exacerbate the already evident over-exploitation of coastal resources and environmental degradation of many coastal habitats. Different sectors often adversely impact each other through conflicting or competitive use of natural resources. Examples of such impacts are as follows:

- Uncontrolled and unregulated discharge of tailings from the Paguna copper mine on Bougainville Island in Papua New Guinea destroyed freshwater fish habits in the Jaba River and also marine habits on the reefs surrounding the mouth of the river.
- In the Western Province of the Solomon Islands logging of tropical forests caused erosion and siltation of rivers. This silt, in turn, was deposited in coastal areas and this adversely impacted the baitfish fishery which was critical to the pole-and-line tuna fishery.
- In Seychelles, inappropriate constructions and navigation channels, destruction of dunelands, unauthorised mining of sand, deforestation, insufficient drainage are eroding coastal areas. Coral reefs are also being degraded by illegal anchoring.
- In Mauritius, coral reefs are being degraded by fishing gear and anchoring, waste from the tourism industry, and sand extraction. Coastal constructions and soil erosion modify the sediment dynamic and thus adversely impact coral reefs.

Because various sectors interactions are often more intense in SIDS than in other countries, integrated planning and management would allow a better co-ordination between sectors so that environmental externalities could be minimized, conflicts between users reduced, and synergies enhanced. More specifically, integrated coastal area management would entail a participatory process of problem and solution identification that will benefit all sectors and users. The process should result in the definition of an agreed strategy, including institutional arrangements that will allow each sector to address interactions and include the necessary measures to enhance production while conserving the resource base. Such a process however, needs to be undertaken gradually and phased as more experience and knowledge are acquired. It will also create more awareness as each sector will increase its capacity to identify the externalities it generates on the environment and quantify and qualify its contribution to the economy, social well-being, and environmental health, including the incurred losses due to environmental degradation.

REFERENCES

Acton, D.F. & Gregorich, L.J., eds. 1995. *The health of our soils. Towards sustainable agriculture in Canada.* Ottawa, Agriculture and Agri-Food Canada.

Hellawell, J.M. 1986. *Biological Indicators of Freshwater Pollution and Environmental Management.* Elsevier Applied Science Publishers, London. 546 p.

Hilborn, R. and Walters, C.J. 1992. *Quantitative Fisheries Stock Assessment.* Routledge, Chapman and Hall.

Pankhurst, C.M., Doube, B.M. & Gupta, V.V.S.R. 1997. *Biological indicators of soil health.* Oxford, UK, CAB International.

Prescott-Allen, R. 1996. *Barometer of Sustainability. What it's for and how to use it?* The World Conservation Union (IUCN), Gland, Switzerland. 25 p.

Müller, R. and Lloyd, R. (eds.). 1994. *Sublethal and Chronic Effects of Pollutants on Freshwater Fish.* FAO and Fishing News Books, Oxford, UK. 371 p.

14

Sustainable Agricultural Intensification

Over the 10 000 years since the Neolithic Age, human population growth has been correlated with intensification in food production systems, primarily agriculture, which has also greatly modified the world's ecosystems, for better or for worse. Given the present and anticipated increases in world population, not to mention current and projected environmental problems and ecological stress from agriculture, further agricultural intensification will be needed. What ethical issues have been raised in connection with past episodes of agricultural intensification? Who can make a difference in directing the course of future intensification, and what ethical responsibilities do these potential actors bear?

Any answer to these questions will be complex. It will involve sophisticated analyses of the historical and natural processes that have influenced and continue to influence episodes of agricultural intensification. It will also involve an account of the social, economic and psychological underpinnings of both individual and collective action, and this account will need to be woven into detailed factual analyses of the land, climatic and genetic endowments in specific places. The focus in this ethics subject is just one element of the problem: the way that ethical concepts and terms can be used to articulate the normative dimension – what *ought* to be the case with respect to human purpose and conduct whenever episodes of intensification occur. Although several examples from the history of intensification are does not attempt to provide an exhaustive or representative survey of ethical issues that have been raised in connection with intensification.

Neither does it offer a general or universally applicable answer to any of the questions above, much less to the comprehensive question, "What ought to be done with respect to intensification?". Instead, the focus is on building capacity for debate and analysis of these ethical questions, with the expectation that specific answers will be the result of more focused inquiries conducted by people with more proximate access to and knowledge of circumstances that are relevant to specific cases. Debate and analysis of ethical issues are difficult for several reasons. In many cultures, ethical norms are deeply embedded in traditions and religious practices, while in others they are thought to be highly personal.

With global issues, the multiplicity of cultural, linguistic and ethical traditions means that one cannot draw upon a common set of terminology or working assumptions. Activities that require coordination to specify or achieve goals or to implement ideas and concepts may be described as "social", "economic" or "political" rather than ethical, leading to a situation in which questions about what "ought to be" are not recognised as ethical ones.

As social science has evolved to analyse activity in these domains, the scientific practice of focusing on data collection and the formulation of refutable hypotheses concerning relationships among data has tended to push aside explicit attention to underlying normative concerns or assumptions. The ethical dimension of such questions can be obscured by implicit and unquestioned assumptions. Each of these factors complicates the articulation, defence and critique of ethical issues. It is therefore useful to undertake planning and analytical exercises in which ethical questions are the primary focus.

The goal here is to provide a basic conceptual framework in ethics that will aid in the articulation and justification of norms for intensification, although this framework will also be applicable in other areas. As already stated, does not endorse a particular set of answers to the ethical questions that arise from intensification. Instead, conceptual tools that allow these questions to be articulated and presented and examples are given for illustrative purposes.

AGRICULTURAL INTENSIFICATION STRATEGIES

Agricultural intensification can be technically defined as an increase in agricultural production per unit of inputs. For practical purposes, intensification occurs when there is an increase in the total volume of agricultural production that results from a higher productivity of inputs, or agricultural production is maintained while certain inputs are decreased.

Intensification that takes the form of increased production is most critical when there is a need to expand the food supply, for example during periods of rapid population growth. Intensification that makes more efficient use of inputs may be more critical when environmental problems or social issues are involved. In either case, changes caused by intensification are to be understood conceptually in contrast to extensive adjustments, which involve increases or decreases in the amount of inputs used.

Historically, the most common and effective extensive adjustment in agricultural production has been to increase or decrease the area of land planted. The contrast between intensification and extensive adjustment is intended to indicate the contrast between two broad strategies that human beings have had for affecting their food supply, rather than concepts applicable to economic or technical analysis of specific cases.

The technical specifications and measurement of intensification or extensive adjustment in any given case are actually quite complex. Changes in the productivity of one input are likely to be accompanied by adjustments in the amount of other inputs. This complexity notwithstanding, there is little doubt that agricultural intensification has been a prerequisite to human civilisation.

The Neolithic technological revolution was built on collecting, concentrating, selecting and harvesting plant and animal species in an organised fashion, with the aim of having more products closer to hand and easier to convert into nutrition. The domestication of farm animals and the development of crops, in the context of ever more productive farming systems, enabled the human population to grow and towns and villages to develop, having governments, laws, trade and economies with specialised employment.

As agricultural production became more efficient, so populations increased. Historians have argued whether higher populations drove technological development, or whether technological development made higher populations possible. Nevertheless, throughout this development, most societies were chronically malnourished, or prey to episodic famine. Relatively high transport costs meant that most societies relied on local production, except when water transport made imports possible. Classical Athens was largely fed from the Crimea, and Rome from Egypt and southern Spain. For more than 95 percent of the history of civilisation, food has been scarce for nearly all people. This has meant low life expectancy, susceptibility to disease and little capacity to face wars, droughts, floods and other human and natural catastrophes.

Food scarcity and social disorder brought about major migrations of people and caused wars and massive cultural disruption. Conversely, when scarcity was relieved, major cultural advances were made. About 1 000 years ago, new varieties of rice were taken to southern China from the Champa Kingdom.

These were not sensitive to photoperiod, and produced two crops a year instead of only one. When these varieties were grown, the intensive techniques that had developed slowly in China to increase productivity per unit of land, although requiring more labour, resulted in dramatic production gains.

The subsequent rice surplus triggered changes across various sectors of China's economy, stimulating the construction of roads, canals, dams, ironworks, grain storage facilities and the production of weapons. For five centuries southern China experienced sustained economic growth and had favourable trade terms for silk, spices and technology with late Mediaeval and early Renaissance Europe.

The agricultural revolution in the early modern United Kingdom greatly increased agricultural productivity. It relied heavily on techniques based on horse power, soil-supporting crop rotations, land drainage and grazing systems, which were developed on the European continent in the very densely populated and often scarcity-ridden Low Countries. When these innovations were applied to larger areas at lower human population densities, a significant surplus was produced, which made the Industrial Revolution possible. At the same time, the harnessing of energy sources became more efficient.

The late eighteenth century saw improved water mills and windmills, the use of sea coals and the development of steam engines. Transport costs fell, making it economically viable, for example, to ship bones long distances for use in fertiliser. Internal combustion and hydropower later greatly increased returns to labour and the availability of products such as nitrogen fertilisers. However, any technology exists within a social and political system.

The way that this British agricultural surplus was generated, controlled and distributed under the restrictive Enclosure Acts forced most of the rural population off the land to serve as industrial labour. The result was high food insecurity and a structural form of urban poverty, where families no longer had access to land or to traditional rights such as commons and gleaning. Yet the growing surpluses allowed greater specialisation, provided capital and cheap labour for industrialisation in Europe and drove the long, violent European nineteenth century. They also resulted in much larger markets in food to supply a burgeoning population, which produced many more goods and services than at any time in previous history. However, the costs paid in human suffering by three or four generations of impoverished families were considerable.

An understanding of the ethical issues involved in intensification can take both a prospective and a retrospective outlook, and ethical standards for evaluating intensification can take either a broad outlook on the general trend of events or a specific focus on the particular responsibilities of key actors.

ETHICS FRAMEWORK FOR AGRICULTURAL INTENSIFICATION

The word "ethics" refers to principles or standards that define behaviour, action or rules for action that is considered to be right, good and proper. A framework for organising the enormous variety of ethical standards that have served in this role throughout human history can be developed from a simple schema of human action. Actors considering or initiating action do so under a set of constraints, represented by the rings encircling the shadow person. These constraints determine which action or behaviour is possible, and are of three kinds. First, some constraints determine the physical universe of possibility. Characters in science fiction novels may be able to dematerialise and transport themselves to other places, but human beings cannot. Constraints that determine the physical limits of possible actions represent *technology*. Second, *law and policy* limit the universe of possible behaviour and action that an actor will consider. Third, individuals and associations constrain their behaviour according to *customary norms* that often lack any legal or official sanction, yet may function very effectively to limit the universe of possible alternatives for action. For example, people from Western societies will spontaneously queue for service, although in most instances this norm lacks legal reinforcement. Together, these three types of constraint make up the *opportunity set*, the class of actions or behaviours that are effectively available to any potential actor.

Eventually, the actor will select one possible course of action from the opportunity set and will engage in conduct. Conduct indicates the behaviour performed, including both physical motions and symbolic or meaningful behaviour. Clearly, the significance of behaviour and the interpretation of the relatedness of multiple acts depend heavily on the broader social context. There may be room for differences of opinion about what, exactly, constitutes conduct in a given instance. For the present purposes, conduct is inclusive of all acts understood as components in an actor's performance of an action.

The term "consequences" here designates especially changes in the health, wealth and well-being of affected parties that are caused by the initial action. As with conduct, there may be differences of opinion about what these consequences are, especially when consequences are indirect or are remote in space and time. Three distinct ways in ethical principles can be developed to determine whether an action is right, good and proper.

First, it is possible to see the ethical validity or correctness of an action as a function of its consequences. Increases in the health, wealth and well-being of people are generally characterised as *benefits*, whereas adverse effects on health, wealth and well-being are characterised as *harms* or *costs*. Right, good or proper actions will tend to be seen as those that have achieved the best balance of benefit and harm relative to other possibilities in the actor's opportunity set.

Second, it is possible to see the ethical validity or correctness of an action in terms of its consistency with an ideal set of constraints. These constraints may be articulated either as *duties* that the actor must discharge, or as *rights* held by others, which the agent must respect. Rights and duties are generally correlated, however, so that if one person has a right, others have a duty to respect it, while having a duty means that others have a right to expect that the duty will be discharged.

Finally, it is possible to see the ethical validity or correctness of an action in terms of conformity to certain types of conduct. Instances or patterns of conduct that are ethically right, good and proper are *virtues*, while those that are wrong, bad or improper are *vices*. This third pattern of ethical evaluation lends itself particularly to expressions of ethical judgement that emphasize the character of the actor, so that not only is the act virtuous, but also the person who reliably acts in virtuous ways.

Utilitarian Model

Agricultural intensification is a process that occurs when individual human beings, communities or organisations take actions of one sort or another. The ethics framework can be applied to the actions of individuals, associations and organised corporate bodies. The framework illustrates three ways in which questions about the justifiability and ethical acceptability of any particular course of conduct might be posed and analysed. Applied to agricultural intensification, these become questions that might be pertinent to the acts of farmers, input suppliers, technology developers, or to any number of public and private agencies whose activities affect the productivity of agricultural inputs. Such questions might be asked by a person or group evaluating their own options and might also be posed as part of a general debate about what actions should be taken by governments, international agencies or, indeed, any actor in the food system.

Since policies are, in fact, actions taken by governments or other organisations, it is possible to evaluate the institution of any policy that would affect intensification in much the same way as one evaluates any ordinary act. The first ethical question is to determine what it is about intensification that makes it a good thing, something to be encouraged or brought about in a particular set of circumstances and, correlatively, what circumstances might make intensification an ethically bad thing.

Consequentialist ethical approaches provide the most straightforward and obvious way to evaluate an entire system of food and fibre production. The consequentialist understands what is right, good and proper to be determined by the impact of the action or policy on health, wealth and wellbeing. Intensification is, in the prototypical case, intended to increase the total amount of food available without increasing the use of inputs.

Since food is material to human life and health, the production of more food can be considered a beneficial impact, especially under the circumstances of food scarcity that have been too typical of human history. As noted, intensification is associated with periods of human population growth. Without a correlative growth in food supplies, food scarcity causes hunger, disease and starvation. Using the framework previously described, many individuals and groups consider options and undertake conduct that has the outcome (consequence) of increased food supply.

The benefits associated with increased food availability provide the elemental argument for intensification, and this argument is consequentialist in its moral logic. In the simple case where new technology or farming methods allow a farmer or landowner to produce more food, consequential reasoning shows why this is ethically a good thing.

As already stated, European agricultural intensification immediately prior to the Industrial Revolution was accomplished not only by applying a package of new production technologies to farming, but also by the Enclosure Acts, which disestablished a system of rights and duties that permitted commoners to live on and farm lands as long as their crops were shared according to an ancient formula.

The framework applies not only to the conduct of individual farmers and landowners, but also to the political activity that led to this policy change. Was enclosure ethically justifiable? The British philosopher John Locke posited the following argument to show that it was:

> ... He that encloses land, and has a greater plenty of the conveniences of life from ten acres, than he could have from an hundred left to nature, may truly be said to give ninety acres to mankind; for his labour now supplies him with provisions out of ten acres, which were but the product of an hundred lying in common.

Here the disestablishment of the old system of commoner rights and duties is portrayed as justified in light of the increased benefits (conveniences) accruing from enclosure. Although this style of thinking is not typical of Locke, the passage implies that any system of rights and privileges is justified, given the efficiency with which it supplies human beings with provisions.

Efficiency is particularly important in the most common form of consequentialism, *utilitarianism*. Utilitarians assume that the values associated with consequences can be quantified to produce a ranking system for all possible courses of action (or *options*) available to an agent. They also assume that the value of benefits and harms can be added and subtracted. Such a ranking system produces a class of optima such that no option in the opportunity set yields greater total value.

According to the utilitarian standard, the right, best and proper action or policy must be a member of this class of optima. This is popularly stated as: Act so as to produce the greatest good for the greatest number of people. Thus, the most efficient approach to producing benefits or avoiding harms is the course of action most thoroughly justified by ethics. Utilitarianism has been an implicit ethical philosophy for agricultural science, which has sought to "make two blades of grass grow where one grew before".

One critical and often overlooked aspect of the utilitarian approach is the need for a complete accounting of costs and benefits. The green revolution involved new seed varieties that were more responsive to nitrogen fertilisers, which are, in most settings, a purchased input. Hence, a simple utilitarian approach weighs the benefits of increased yields against the costs of seeds and fertiliser. If benefits outweigh costs, the green revolution is justified. Yet other shifts accompanied the new technology, and many critiques of the green revolution can be articulated entirely within the framework of a utilitarian/consequentialist ethic. Within the first decade of the green revolution of the 1970s, which led to aggregate increases in rice production across Asia, large-scale insect pest outbreaks and plant disease epidemics destabilised food production, supplies and prices.

Only after three boom and bust cycles and pest outbreaks did governments begin to move away from simplified, centralised pest control policies that relied on insecticides and vertical host plant resistance towards decentralised integrated pest management that built on local ecological processes to realise production potential. With the concentration of farm animal processing facilities, the chances of large-scale epizootics occurring increase exponentially because of the more extensive movements of animals between pasturage, feedlots and abattoirs, and contacts with animal offal and excrement. Feeding livestock with products derived from their own species creates routes for infection by diseases associated with prions.

Short-rotation forest plantations increase wood production but at the same time increase vulnerability to specialised pests and diseases. Fast-growing dwarf coconut varieties increase short-term yields but are more frequently at risk from diseases previously found in limited geographic areas. In fisheries and aquatic production systems, exotic species are commonly introduced. These often initially increase total production, but can unexpectedly change trophic relationships and disrupt ecosystems, as did the Nile perch in Lake Victoria. Intensive salmon hatcheries have been criticised for reducing the genetic adaptability of natural populations. These problems testify to the need for completeness in thinking through the costs and benefits of intensification.

Rights-based Model

John Locke's rationale for enclosure makes no mention of rights, but the language of rights might be used to develop a rationale for arguing that efficiency-seeking public policies (such as enclosure) are ethically wrong in certain circumstances. Simply stated, policies are unjustified if the changes they endorse violate or override important rights. Among the rights that might be violated are property or occupancy rights, if enclosure forces people who hold such rights off the land, or subsistence rights, if the effect of enclosure is to put some people in a position where their rights to food and shelter are not met. Yet the ethical principles behind these rights are complex.

A rights-based approach to ethics proceeds by stipulating or deriving a set of basic rights and duties that agents must perform without regard to the consequences that might arise in any particular case. A rights-based approach to intensification is more concerned as to whether the actions that result in higher food production are consistent with these rights and duties than in their eventual effect on human welfare. Several methods have been put forward for identifying these rights and duties.

The German philosopher Immanuel Kant developed complex philosophical arguments for a master principle he called the *categorical imperative*: never act in a way that treats another person solely as a means to an end. For a Kantian, rights and duties can be traced back to this principle, which tells us always to respect other people's capacity for freely choosing their life plans. The problem that a Kantian would have with Locke's argument for enclosure is that it seems to treat the commoners (whose rights are violated) simply as a means to the larger end of increasing food production.

Kant's deontology came at the end of a longer tradition of natural law theory, which held that the basic rights and duties forming the main content of the moral law are evident to any rational person, and have hence invested great effort into rational argument for certain approaches to the configuration of rights. Kant argued that every human being wants to be treated as a free, autonomous agent, and that consistency requires people to treat others in the same way. For other philosophers, going back to Thomas Hobbes, rationality was understood simply as enlightened self-interest.

The purpose of rights is to protect people's ability to act on their own enlightened self-interest. Hobbes believed that any rational person would agree to live in accordance with certain rights and duties, if only they could have a reliable expectation that everyone else would do the same. This approach is sometimes characterised as contractualist in light of the way that rights and duties are described as being grounded in an implicit agreement among all members of society.

While a Kantian might argue that enclosure treated commoners as a mere means to a larger social goal, a contractualist might say that enclosure violated the social contract. In either case, the problem with enclosure from a rights-based perspective was that commoners had a right to use these lands. As such, any plan to exclude them from the land could not be justified unless it included some provision for obtaining their agreement. This would require involvement of the commoners at some stage of the enclosure process, and their involvement would need to be such that each rights holder had the opportunity to give or withhold their agreement voluntarily to the plan.

Perhaps they could be convinced that they would be better off, or perhaps they would be enticed to accede to such use of the land in exchange for compensation. These details might vary considerably on a case-by-case basis, but what is critical from a rights-based perspective is that respect for the individuals affected by enclosure requires that they be accorded a role in the intensification process that is fully consistent with their rights.

Either version of rights theory provides a starting-point for questioning whether intensification is good simply because it produces more benefits than costs.

Either approach to rights implies that the exclusion of commoners from decisionmaking is unacceptable, but a Kantian might in addition note that the effect of enclosure was radically to impoverish commoners and their descendants. For a Kantian, such conditions of poverty make it impossible for a person to exercise rational free will; the circumstances of need are so great that people in dire poverty are effectively coerced into enduring humiliation and deprivation.

As such, key subsistence rights, including the right to food, become minimal conditions that must be met if all people are to be treated with the moral respect to which they are due. Thus, any situation in which people are so poor that they cannot freely exercise their innately human capacity to choose a life plan involves an ethical wrong. Methods of intensification that place people in such circumstances cannot be endorsed from a rights-based perspective.

Issues in Utilitarian and Rights-based Approach

The basic tension between utilitarian consequentialism, on the one hand, and rights-based ethics, on the other, underlies many issues associated with agricultural intensification. For decades, researchers from developed countries have harvested germplasm from farmers and local markets in the developing world. These researchers have used the germplasm in breeding programmes to develop higher-yielding varieties as well as searching for other valuable genetic traits.

Many of those who collected seeds were shocked when critics suggested that their work failed to respect the rights of people from the developing world. From a utilitarian viewpoint, the increased yields of new varieties more than justified the collection of germplasm, and researchers saw no ethical issue in using seeds they had collected this way. However, critics asserted that researchers had failed to show proper respect for the rights of indigenous farmers whose forebears had saved seed for centuries.

Purchasing seed in village markets, critics affirm, gives the buyer an implied right to use the commodity good for food, or possibly for replanting, but farmers could not be interpreted to have given up rights to further development of their germplasm without a careful and explicit process to inform them of its true value and to ensure that they had given consent. Some critics argue that because of the collective and collaborative nature of seed development in traditional agriculture, only someone who represents the collective interests of all growers would be in a position to undertake such a negotiation.

Today, some opponents of genetically engineered crops argue that individual consumers should not be forced to eat these crops against their will. Advocates of genetically engineered crops see them as a safe and effective tool for increasing the efficiency of farm production and believe that their opponents' claims are an unjustified barrier to adopting them.

Irrespective of the safety or risks associated with genetically modified (GM) crops, opponents are claiming that the companies promoting these technologies have placed food consumers in a position where they have no opportunity to reject them. Rather than rebut the claim that consumer rights are at stake on its merits, advocates have often argued that if the crops are safe, then consumers have no basis for rejecting them, since rejecting GM crops is an action with real costs but no real benefit. Thus, opponents try to meet a rights claim with a utilitarian argument, and the ethical issues fail to be enjoined.

A similar point of tension arises in debates over the green revolution technologies. When higher yields are associated with purchased inputs, those with access to capital will have an advantage over those without. One can interpret losses experienced by poorer farmers as additional costs and weigh these, too, against benefits from higher yields.

The study *New seeds for poor people* by Michael Lipton and Richard Longhurst is a particularly exhaustive and theoretically sophisticated attempt to assess the green revolution varieties from a utilitarian standpoint. The authors conclude that, over time, these technologies have been, on balance, beneficial to poor people. However, these arguments do not necessarily address concerns that rights may have been violated, or that cultural traditions may have been lost as a result of green revolution strategies.

It is possible to see any transition from a situation in which people can feed themselves and meet their needs to one in which they cannot as a violation of their rights. Even if such transitions have benefits that outweigh the cost, they would not be seen as justified if there are even a few individuals whose rights to subsistence are jeopardised as a result of the changes. This is only a cursory introduction to the way that rights arguments should figure in an overall evaluation of the green revolution. Yet one of the problems that has arisen in debates over the retrospective impact of green revolution varieties is that those who draw their ethical norms from utilitarian thinking seem to be ignoring ethical claims that draw upon the language of rights. In doing this, people create an impression of, at best, insensitivity to the full range of ethical concerns relevant to intensification and, at worst, arrogant dismissal of arguments that are inconvenient for the case they wish to present.

When people who hold influential positions for future attempts to meet food needs dismiss alternative arguments in this way, they engage in a use of power that is itself ethically questionable. As such, it would be valuable to launch retrospective studies that make an explicit attempt to acknowledge the full range of concerns that have been or might be brought to bear on evaluating green revolution intensification, in part as preparation for more open and informed debates on the questions that must be addressed if world food needs are to be met in the twenty-first century. These cases are complex and deserve a more careful analysis than these summary statements can provide. The point here is simply to note that, in both cases, a utilitarian/consequentialist rationale is met with counterclaims that assert rights. Someone assuming that utilitarian models are appropriate will respond to these assertions with further arguments reciting the costs and benefits of alternative arrangements, but to the extent that these rights are thought to be moral rights, rights protecting the dignity of involved parties, recitation of further costs and benefits will simply miss the point.

What is being claimed is that there is a need to respect affected parties by involving them fully and non-coercively in the process of intensification. From the perspective of rights-based ethics no recitation of costs and benefits will justify a failure to do this; what is needed is a justification that addresses how the process of intensification influences the freedom of all persons affected by it. If freedom is being constrained, it must be shown that this constraint is justified, perhaps because it is required in order to show proper respect for the rights of less fortunate people, or perhaps because the people agree to the constraint of their own free will.

Virtue-based Model

The utilitarian/consequentialist and rights-based approaches are both theoretically well-developed articulations of ideas used every day by people who are attempting to determine the correct or proper course of action. However, it may be more typical for people to associate ethics with less systematic ways of thinking. For example, many people may address ethical questions by asking themselves how some exemplary person would act in a given situation. This exemplary person might be a family member, a revered and respected member of the community, a religious leader or perhaps even a person from legend or history who is not even known as a real flesh-and-blood individual. In any case, one draws upon a mental image of how a good person would act in the situation at hand. By doing this, one understands the ethics of the situation in terms that refer directly to the conduct that is being performed, rather than either the rights-based constraints or the consequences of the action.

The philosophical possibilities for developing this general approach to ethics are numerous. The Greek philosopher Aristotle did so by developing a catalogue of both positive and negative exemplary types of conduct, or virtues and vices, respectively. He theorised that all human tendencies had appropriate forms of expression in moderation but could become vicious if not held in check. A good person is one who strikes the mean position among these tendencies. Moderation itself thus became the sovereign virtue for Aristotle and many of his followers. Today, many types of ethical theory that emphasize conduct, rather than rights and duties, on the one hand, or consequences, on the other, are referred to as Aristotelean or, more accurately, virtue theories. Although some philosophers have believed that virtues and community traditions are relatively undeveloped sources of ethical insight, better articulated by emphasizing rights and duties on the one hand, or consequences on the other, others have argued that this domain of ethical ideas or source of ethical insight cannot be eliminated.

One contemporary school of ethics that stresses the social roots of exemplary conduct is called *communitarianism*. Here, the articulation of ethical norms and standards is likely to call attention to the norms, practices, traditions and institutions that are particularly characteristic of and valued by a particular community, rather than particular virtues and vices. Contemporary communitarians emphasize the need to have a social environment, a form of community life, that will give rise to exemplary conduct and that will allow people to appreciate the ways of life that such forms exemplify.

Agricultural ways of life have figured prominently in some of the most influential articulations of virtue and vice. The ancient Greeks themselves developed a form of agriculture based upon rough terrain, varying soil types and a Mediterranean climate. Their farms were a diverse mixture of grain production and pastoral livestock, but with heavy reliance on tree and vine crops. The mix of crops and long growing season provided steady work for fairly small households all year round, while the trees and vines involved lifetime investments for smallholders. Military historian Victor Davis Hanson argues that this pattern of agriculture gave rise to unique forms of military organisation and tactics as well as the political culture of the city-state.

The relatively large proportion of the population controlling property and the nature of their stake in the land made them both fierce defenders of egalitarian political forms and equally fierce warriors who could be relied upon for phalanx manoeuvres requiring discipline and loyalty. These character traits, so critical to the success of Greek city-states as political and military entities, were thought to emerge naturally in a farming population of smallholders. In contrast, the large-scale plantation-style irrigated agriculture common among the Greeks' military rivals relied on stratified societies of slaves and masters who did not develop the requisite virtues.

REFERENCES

Carrol, C. R., J.H. Vandermeer and P.M. Rosset. 1990. *Agroecology*. McGraw Hill Publishing Company, New York.

Horne, James E. and Maura McDermott. 2001. *The Next Green Revolution: Essential Steps to a Healthy, Sustainable Agriculture*. Food Products Press, an imprint of The Haworth Press, Binghamton, NY. 312 p.

Pretty, J.N. 1994. *Regenerating Agriculture*. Earthscan Publications Ltd., London.

Savory, Allan, with Jody Butterfield. 1999. *Holistic Management*. Island Press. Washington, D.C. 616 p.

Suthipradit, S. et.al. (eds) "Sustainable Agriculture: Possibility and Direction" Proceedings of the 2nd Asia-Pacific Conference on Sustainable Agriculture 18-20 October 1999, Phitsanulok, Thailand.

Thies, C. and T. Tscharntke. 1999. "Landscape structure, and biological control in agroecosystems". *Science* 285: 893-895.

Thompson, Dick. 1997. *Alternatives in Agriculture*: 1996 Report. Thompson On-Farm Research and the Wallace Institute. p. 3-4.

Vandermeer, J. 1989. *The ecology of intercropping*. Cambridge University Press, Cambridge.

15

Environment Education for Sustainable Agriculture

One of the important resolutions of the Rio de Janeiro in 1992 was Agenda 21, which specified the need to increase environmental awareness and undertake specific public education programmes to influence positive attitudes and appropriate behaviour of various segments of the population towards sound and responsible environment and natural resources management as well as sustainable development. Several years have passed since the Rio Earth Summit, where most country leaders had agreed to take action in preserving the environment through communication, education and training programmes. Few such activities have been implemented, especially for the rural population, despite the fact that many environment factors directly and seriously affect agricultural productivity and the well-being of farm families.

Nevertheless, in the last couple of years an increasing number of decision-makers in the agricultural sector are interested in mainstreaming environment conservation and natural resource management issues and concerns into broad-based agricultural policy formulation, programme planning, extension, education and training activities. In response to such a trend and in view of the above-mentioned needs, FAO established a new department in 1995 called Sustainable Development and intensified efforts in promoting environment education and training among its member countries. Through its Extension, Education and Communication Service (SDRE), FAO initiated field-level environment education and training activities in close collaboration with strategic partner institutions in a number of Asian countries.

ENVIRONMENT EDUCATION THROUGH AGRICULTURAL EXTENSION TRAINING

One of the strategic approaches in disseminating and sharing environment issues or concerns with the public is by "piggy-backing" relevant messages through existing communication channels which have large and regular clientele, in an institutionalised and sustainable manner. In most developing countries, the majority of rural families have relied on field agricultural workers for information and advice on agricultural and rural development. However, in many countries, agricultural extension service is often weak and ineffective, extension workers are not well trained and have a narrow scope of expertise mainly limited to agricultural production technologies. As many environment issues are directly related to sustainable agricultural development, there is now a need and a new opportunity to revitalise and strengthen such a service by mainstreaming environment education through agricultural extension training.

The participatory-approach and peer-learning methods employed in planning, implementing and evaluating EET of agricultural personnel and the institutional and professional networking strategies that have facilitated close collaboration and sharing of know-how and experiences on Environment Education Training Modules (EETM) planning, development, testing and utilisation among eight institutions from six Asian countries - China, Bangladesh, Thailand, Indonesia, the Philippines and Malaysia - are explained. Concrete results from these two-year development cooperation activities are demonstrated and evidence for sustainability and institutionalisation of EET activities in these countries are discussed. The process and strategies of a similar programme sponsored by FAO and UNFPA, which incorporates population and environment education concerns into agricultural extension training in 12 countries in Africa, the Near East and Latin America, will also be discussed and used as comparison.

Strategies, Process and Context

Needs-based, demand-driven diagnostic planning

The integration and development of EET through agricultural extension programmes started as early as 1987 in Indonesia through an FAO technical cooperation project requested by the State Minister's Office of Population and Environment, in collaboration with the MOA, to design and test a methodology for involving field agricultural extension workers in assessing and reporting environment problems which negatively affected farm management and productivity in two pilot provinces - Grobogan District, Central Java, and Central Lombok District, Lombok.

While the FAO project TCP/INS/4514 on Community Extension and Training in Environment Conservation and Protection was successful in developing a strategy and method for undertaking environment surveillance and designing needs-based, problem-solving, environment activities with active participation of agricultural extension workers, concerned farm families and local leaders, it did not include institution building, staff capacity development or training components. Extension workers served as facilitators in encouraging a two-way flow of information and opening up communication channels between relevant government officials, local leaders and concerned farm families. However, such EET activities were undertaken on an ad hoc, sporadic basis and not integrated into planned or regular agricultural extension and training programmes. Hence they were not sustainable, nor were they institutionalised.

The evaluation results from this two-year project by independent researchers from the University of Indonesia and subsequent discussions among FAO/SDRE and MOA and State Ministry of Population and Environment officials concluded that expansion of project activities to other areas would require: (1) wider involvement of agricultural extension workers in an institutionalised manner, (2) integration or incorporation of environment education into regular, ongoing, relevant agricultural extension programmes and (3) systematic and professional quality training of extension workers on relevant environment education issues, strategies and methods by authorised agricultural training institution(s). The Agency for Agricultural Education and Training (AAET), Indonesia, whose staff were involved in the project, felt there was a critical need to include environment education in agricultural training curricula, and concluded that a systematic and institutionalised effort to develop its staff capabilities in undertaking cost-effective EET activities would be given a high priority.

Partnership with strategic training institutions

In view of these conclusions, possibilities for training agricultural extension workers on environment education in Indonesia were studied and explored. Instead of having a conventional approach whereby an FAO project with its world-class experts and foreign consultants trains extension workers on environment education, FAO/SDRE with its limited resources agreed to collaborate with AAET, an agricultural training institution which provides in-service training to all the 33 000 Indonesian extension workers, in developing institutional and staff capabilities to master the process and methods of designing and utilising well-planned and pre-tested EET curricula, modules and learning materials.

Through a simple Letter of Agreement (LOA) between FAO/SDRE and AAET a collaborative activity in environment education through agricultural extension training was started in 1992. With an initial funding of $10 000 and some technical support from FAO/SDRE, AAET was responsible for designing a participatory process and method involving its master trainers, selected extension workers, environment specialists, local leaders and other relevant stakeholders. The task was to develop, test and produce a 15-17 hour EETM for use by its trainers in training extension workers, which would be incorporated into AAET's inservice training curriculum. Subsequent LOAs were made for follow-up phases of the environment education training process, such as training of master trainers in EETM utilisation, pilot trials on EET of field extension workers and packaging and production of prototype EETM materials; FAO also contributed an average of $10 000 - $12 500 for each phase. However, expenditures for scaling up activities such as reproduction of EETM materials and of training large numbers of field extension workers, have now been covered by AAET through their own regular budget.

The training institutions participating in EET activities in other countries are also those that provide training to most of the agricultural extension and other field outreach workers. Such institutions are the key agricultural training centers in their respective countries or have the official mandate to carry out in-service training for agricultural personnel. Furthermore, these selected institutions perceive that their mandate is compatible with and supportive of environment education goals. The incorporation of environment education into their existing training curriculum and extension programme is expected to generate synergy that will facilitate EET programme sustainability and institutionalisation.

Participatory planning

Based on the encouraging results and positive experiences of AAET's EET activities in Indonesia, a participatory environment education programme through agricultural training was designed for interested institutions in the Asia region. AAET's basic (1st level) EETM was translated into English by the Agricultural University of Malaysia (UPM) staff members who had also participated in the pretesting of the EETM.

FAO/SDRE sponsored a regional planning workshop for EET in Kuala Lumpur, Malaysia in June 1994, which was organised by UPM. Twenty-three participants from ten countries attended the workshop. These workshop participants represented public-sector training and extension agencies, NGOs, universities, international organisations and multidisciplinary specialisations ranging from environment science and management, agriculture, economics, public policy, education, training, communication and rural extension. They discussed the needs, problems and strategies for incorporating environment education through agricultural extension and training. The methods of EET activities in Indonesia were shared and the English version of the Indonesian EETM was presented as a prototype to determine its appropriateness for possible adaptation, testing and use as a generic model. Workshop participants suggested recommendations for incorporation of environment education into extension and training programmes and several of them submitted proposals for EET activities by their institutions.

Workshop participants discussed and agreed on EET's conceptual framework, strategy and methods, operational processes and implementation procedures. Such EET standard operating procedures or guidelines included participatory mechanisms which will ensure the relevance and client-orientation of EET programmes, and provide horizontal knowledge interchange and experience sharing among institutions and EET planners within a country or region, or globally. Eight institutions from six Asian countries decided to undertake EET activities, initially supported by FAO/SDRE with small grants ranging from $7 000 to $10 000. Several workshop participants who are not undertaking EET activities agreed to join a panel of non-implementing EET resource persons who could provide advisory, review, monitoring and trouble-shooting services. This network of resource persons has proved critical in providing participatory consultations, peer learning and reviews, quality assurance and standards, as will be discussed later.

Franchising and wholesaling of training

The predominant strategy of this EET programme is the aim of "franchising" of EET activities to qualified and interested training institutions, instead of having FAO/SDRE or FAO projects conduct EET activities in an ad hoc manner. The preoccupation with the "number game" in terms of quantity of training participants should be shifted to developing partner institutions as high-quality training multiplier agencies. Instead of "retailing" such EET courses by itself, FAO/SDRE's limited resources are used for developing the capacity of training institutions and their staff in planning, designing and implementing cost-effective EET activities on their own.

The concept of EET "wholesaling" to training institutions is an approach which focuses on facilitating sustainability and institutionalisation rather than initiating a donor-driven external EET programme dependent on funding. Furthermore, through "franchising" of EET methods, operational processes and implementation procedures to interested training institutions, ETT expertise is transferred and quality and standards are assured. More important, it transfers institutional responsibility for advocating further "wholesaling" and/or "retailing" of EET "product lines" based on local needs and problems.

The EET activities, therefore, emphasised two important aspects: (1) providing opportunities for EET knowledge sharing and partnerships among agencies countries through regular workshops, technical consultations and review meetings at institution, country and regional/global levels, and (2) providing technical support for participatory staff training in EET process and methods such as needs assessment, training module development and pretesting, multimedia materials packaging and production, training of trainers and training module utilisation for training of extension workers.

EET activities are oriented to ensuring that participating training institutions and agricultural trainers master the process and methodology for integrating environment education into agricultural training and/or extension programmes, rather than on specific technical environment subjects. It is critical that EET activities be carried out by a multidisciplinary team, with environment specialists providing relevant technical inputs.

Inter-agency consultations, technical review meetings and contents validation seminars are an integral part of the participatory EET process and methods suggested during the 1994 EET planning workshop in Kuala Lumpur.

Other strategies geared to facilitating the "franchising" of EET activities in training institutions include: (1) initiating EET needs assessment, whose results are used for existing training curriculum reviews and reforms, (2) participatory EET curriculum design and training materials development and testing, (3) training of master trainers in EETM utilisation, (4) piloting trials of EET activities for extension workers, and (5) obtaining EETM contents validation and EET legitimisation as an integral part of the existing training curriculum or extension programme.

The EET process and method guidelines were developed with the purpose of increasing stakeholdership among concerned environment education trainers, extension or outreach workers, local leaders and policy advocates. Methods to solicit active participation and involvement of relevant persons and institutions willing to "buy-in" to the EET activities are of critical importance. Participatory training needs assessment and problem identification methods can facilitate strategic planning for designing a demand-driven, problem-solving, environment education programme. Participation of key stakeholders at all stages of EET planning, development, implementation and monitoring is critical for transferring technical and management skills and programme responsibilities.

Quality assurance and standards

There is a difference between "wholesaling" and "franchising" of training. In the context of EET activities, the concept of "franchising" is more suitable as it emphasizes the need for training quality assurance. Concern for quality is critical for participatory EET activities to be carried out by institutions in various countries which will determine and prioritise EET contents based on their own training needs. Ensuring high quality of EET is no easy task. However, it can be facilitated by planning and application of educational methodology tools such as training needs assessment, curriculum development, instructional design, learning methods selection, contents packaging, pretesting, training of trainers and evaluation.

During the 1994 EET planning workshop in Kuala Lumpur, the participants developed generic EET process and methodology guidelines which have since been used in planning and implementing EET activities. The guidelines provided step-by-step standard operational procedures for EET, including suggestions for the development of EETM. Such guidelines are continuously revised and improved following results and experiences of participating institutions and the comments of the EET network members shared during the annual EET regional workshops, for example the 1995 Bali workshop, the 1996 Beijing workshop and various national workshops and consultative meetings.

Participatory peer learning

Institutions participating in the EET programme and agreed on specific EET objectives, measurable outcome indicators, planned activities, an implementation time-frame and resource requirements. FAO/SDRE provides networking opportunities for participating institutions and EET network members, by convening a regional EET workshop at least once a year to facilitate participatory peer learning and review, monitoring of progress and accomplishments and obtain suggestions for improvements at critical stages/phases in the EET process.

The 1994 Kuala Lumpur workshop focused on EET programme strategic planning and participatory development of EET process and methods. The 1995 Bali workshop reviewed the progress and accomplishments of EET module development and testing, discussed EETM improvements, and considered plans for the training of master trainers. The 1996 Beijing workshop discussed the results of EETM utilisation by trained trainers, considered plans for training of extension or outreach workers, proposed improved strategies for EET institutionalisation and replications/expansion for different environment education subjects or advanced topics. Such technical review meetings and consultations at country and institutional levels are encouraged and included in the EET guidelines or standard operating procedures. Through such consultation and assessment among EET network members using firsthand experience, EET quality is constantly scrutinised and improved.

Process documentation

While summative evaluation or impact assessment is an important tool to demonstrate the effectiveness of EET activities, it may not be adequate for providing explanations and insights into the process and methodologies employed in undertaking such activities. It is useful to conduct a process documentation which points out critical issues and decision-making requirements in undertaking EET activities as well as their background or circumstances. Through chronological description and analyses of successful or less successful decision-making during planning, implementation and management of EET activities, important lessons can be learned, and technical and management improvements can be suggested for future replications and expansion of similar activities.

To facilitate comparative analysis of EET activities in various countries, a standardised set of performance, outcome and impact indicators was developed and has been used to collect the necessary data.

Finally, it should be noted that it is important to commit adequate resources to complete the "last-mile" tasks in consolidating, summarising and disseminating the process, methods, results and lessons from EET activities in a user-friendly, attractive and captivating manner, especially aimed at relevant policy and decision-makers, for further improvement, expansion and replications of EET activities world wide.

EET ACTIVITIES

In EET activities conducted by eight participating institutions in six Asian countries, an analysis of the interplay among sustainable agricultural development, rapid population growth and environment deterioration issues was carried out. Taking into consideration the results of training needs assessment, an inventory-analysis of relevant environment education messages and agricultural extension and training contents was undertaken. The purpose of such an exercise was to identify suitable entry points for the integration of environment issues into agricultural extension training contents. Mainstreaming environment education into agricultural extension and training programmes requires the active collaboration of concerned institutions and line agencies. An inter-agency and multi-sectoral policy advisory committee and as a technical steering committee were established to coordinate EET activities.

In communicating environment issues to rural populations, six participating institutions are utilising agricultural training centres which train extension workers or rural outreach workers as the main channel for environment education. The main beneficiaries of these EET activities include master trainers of concerned training institutions, trainers of extension and outreach workers, field workers of public extension service and selected farmers/community leaders. Two other participating institutions have different target beneficiaries. In China, the Centre for Integrated Agricultural Development (CIAD) uses the Central Agricultural Managerial Official College (i.e. Senior Staff Command College) to lobby high-level agricultural policy and decision-makers by advocating the need for, and importance of, environment management policy and education issues for rural populations. The International Institute of Rural Reconstruction (IIRR) in the Philippines is developing and conducting EET activities for rural outreach workers of NGOs in the Philippines. In undertaking EET activities, strategic alliances have been forged among partner institutions such as public sector agricultural extension and training institutions, environment management agencies, rural development training and adult education institutions, NGOs, community based development agencies and multi-media development and training centers.

Environment Education and Training Modules

All participating institutions had conducted EET needs assessment whose results were used as strategic inputs for designing the EET curriculum and developing EETM ranging from 15-25 instructional hours. These EETMs had been pretested and presented for technical review and contents validation in at least two national workshops and two inter-country/regional workshops.

Training of Trainers and Extension Workers

Training of master trainers in the utilisation of EETM had been completed by all participating training institutions where at least two batches of 30-60 master trainers had been trained. Each trained master trainer is in the process of training another 10-12 trainers. It is expected that 500 environment education trainers will be available in the six participating countries. Each of them was supposed to train at least two batches of 20-25 extension workers as part of their regular training programmes by the end of 1997 or at the latest by mid-1998.

EET Institutionalisation Process

EETMs have been reproduced for wide distribution to potential trainers and interested users in all participating institutions. In Indonesia, Bangladesh, China (CIAD and DOE/ Ministry of Agriculture), the Philippines (UPLB) and Malaysia, the EETMs have been personally endorsed by the highest policy-makers and officially adopted for in-service training of agricultural extension workers. In Indonesia, the EETM has been adapted and utilised by several NGOs and a local government institution (East Java Province) and the AAET has developed an advanced-level EETM module and trained additional trainers and extension workers, using their own resources. In Malaysia, Bangladesh and China, training of extension workers was carried out without FAO/SDRE support. In the Philippines, additional funding from other donor agencies had been obtained to cost share EET activities of the UPLB consortium and IIRR. In Thailand, the Continuing Education Center (CEC) of the Asian Institute of Technology (AIT) is applying EET process and methods and utilising selected parts of the eight EETMs for its own environment-related training course.

Regional EET Workshops

Three regional EET workshops had been convened by 1996. The 1994 Kuala Lumpur workshop focused on strategic programme planning and participatory development of EET processes and methods. The 1995 Bali workshop reviewed the progress and accomplishments of EET module development and testing, discussed EETM improvements and considered plans for training of master trainers. The 1996 Beijing workshop discussed the results of EETM utilisation by trained trainers, considered plans for training of extension or outreach workers, proposed improved strategies for EET institutionalisation and replications for different environment education subjects or advanced-level topics. During these workshops, new EET network members were invited to join, including participants and resource persons from Nepal, Ethiopia, Malawi, Kenya and Egypt.

EET in Agricultural Sector

The incorporation of EET to the agricultural sector signals an important implication for the direct implementation of Agenda 21 activities at the grassroots level. One of the sections in the Agenda 21 document of the 1992 UNCED's Earth Summit in Rio de Janeiro focused on the conservation and management of resources for development which contained the issue of sustainable agriculture. This issue addressed one of the most fundamental tasks of global sustainable development to provide adequate food for an expanding world population and at the same time protecting the environment. With these tasks, the agricultural sector now needs to be concerned not only with increasing agricultural production but also ensuring that the environment is protected for long-term sustainability.

EETM Development Process

The EETM development process followed by the agencies/institutions in the six countries may be divided into five phases of activities:

a) Phase 1: initial start-up activities;

b) Phase 2: first EETM regional workshop;

c) Phase 3: EETM development activities;

d) Phase 4: second EETM regional workshop;

e) Phase 5: third EETM regional workshop.

Phase 1 activities: initial start-up activities

Through the initiatives of FAO, the integration and development of EET through agricultural extension programme started in Indonesia in 1987. A pilot project was then implemented to train extension workers in environment conservation management and practices which eventually led to the development of the TM on environment conservation for agricultural extension workers. In 1992, the AAET, MOA, Indonesia, with the assistance of FAO undertook to develop an EETM designed to be used as a basic course in natural resources and environment. The successful case experience of this pilot project prompted FAO to share and duplicate this experience with other countries in the region through a series of planned workshops.

Phase 2 activities: first EETM regional workshop

The development process of the Indonesian EETM was documented and the prototype materials presented at the first regional workshop on environment education through agricultural extension, organised by the Agricultural University Malaysia or Universiti Pertanian Malaysia (UPM), Malaysia and sponsored by FAO, held in Kuala Lumpur, Malaysia from 20 to 25 June 1994. Participants for this workshop were selected from various prospective institutions/agencies in different countries which planned to incorporate environment education in their agricultural extension training system. The activities of the EET development process, undertaken by Indonesia and the prototype materials produced were presented as case materials and discussed during the workshop. These provided the framework and guidelines for such activities to be undertaken by other countries/agencies in the region.

The major outputs of this workshop were (a) a list of recommendations on the critical environment issues and identified priority target beneficiaries, and (b) review of the processes and activities to develop a prototype EETM. There was strong emphasis on the need for a more systematic and holistic approach in dealing with environmental issues, especially through training. The workshop advocated that a more concerted effort be made among the various organisations and expert groups to form strategic alliances to address the most common critical environmental concerns. In addition, the agricultural extension, education and training institutions should explore more innovative methods in their training programmes and have the task of extension to cover issues concerning the environment.

At the conclusion of this workshop, seven agricultural/rural development agencies from five different countries submitted proposals to develop EETM for training different priority target beneficiaries. Various strategies to develop the EETM were used by these agencies; some of them worked in partnership with other in-service training institutions, formed multi-agency group or undertook the project independently; all aimed to optimise the effectiveness from reaching the priority target beneficiaries. These agencies included both government organisations and non-government organisations (NGOs) and they target a range of beneficiaries including the farmers, extension workers and trainers as well as planners and policy/decision-makers. The contents of the proposed modules to be developed by the agencies/organisations reflected the most critical needs of the various target beneficiaries and included both technical contents and policy guidelines as well as advocated environmental issues and concerns relevant to the needs in current situations.

Phase 3 activities: EETM development activities

Having participated and submitted the initial project proposals at the first regional workshop, each of the agencies/institutions from the various countries is now better equipped to undertake the tasks of implementing the EETM development activities following a more systematic and integrated approach. These projects, upon approval by FAO were each provided with a small amount of seed money, ranging from US$5 000-US$10 000 to undertake the planned activities. The proposals submitted during the workshop reflected only spontaneous reactions to their needs during the workshop. The critical environmental issues and concerns, the target beneficiaries and the strategies to be adopted in order to operationalise the activities for the EETM development process were further investigated and validated by the authorities concerned in the home countries. These were linked to the strategies of partnership, consortium or individual approach to undertake the EETM project. For example, in Malaysia, the UPM partnership with two other agricultural training institutions, the MOA and the Muda Agricultural Development Authority (MADA) was important in promoting environment education through agricultural training as almost all the trained agricultural human resources in the country were from these three institutions.

The context of policy advocacy for environment education was important in China and was therefore incorporated in the training of agriculture and rural development planners and local leaders. On the basis of the prominent status of the NGOs in the Philippines, environment education therefore formed part of the training programme of their rural outreach workers. In the other countries, Bangladesh and Thailand, the Department of Agricultural Extension (DAE) played a key role in information and technology dissemination to the rural communities, hence the department undertook the EETM project on an individual basis. The key objective of the various strategies used in the different countries and their collaborative efforts was to enhance the optimal dissemination of knowledge and skills on the critical environmental issues to the target beneficiaries.

The development of the EETM undertaken by these agencies was guided by the recommendations proposed at the first regional workshop. All the agencies/institutions therefore followed the EETM development process based on the framework and lessons learned from the Indonesian experiences. Thus, over the following 14 months, from August 1994, each of the teams of module writers from the various agencies/institutions worked on their prototype EETM based on these guidelines.

Phase 4 activities: second EETM regional workshop

After the given time frame for each of the agencies/institutions to prepare the prototype EETM, a second regional workshop was organised to: (a) review the progress and assess the results of the EETM development activities; (b) share the lessons learned and suggest further improvements in the EETM development process and methods; (c) plan the utilisation of the EETM in a sustainable and institutionalised manner. The five-day workshop, organised by the AAET and sponsored by FAO was held in Bali, Indonesia from 16 to 20 October 1995 and attended by 21 participants, the majority of them having also participated in the first regional workshop. These participants represented the core group of professionals who were considered as EETM network members and were directly involved in the EETM development activities of their respective agencies/ institutions.

The workshop provided tremendous opportunities for the agencies/institutions to share experiences and learn from each other the best practices to be adopted and the mistakes to be avoided. In addition, the experiences and lessons learned from other parallel activities on PETM development were also discussed and highlighted. Prior to this workshop in Bali, a regional workshop for the PETM development project had already been conducted in Chiang Mai, Thailand in March 1995 and attended by 27 participants from 15 countries.

The major outcomes of this second regional workshop were three important recommendations on the strategies and action for improving EETM development made at the conclusion of the workshop. These included (i) the principles to follow in a modular approach to participatory training, (ii) the basic sequential steps in the module development process, and (iii) suggestions for cost-effective utilisation and sustainable institutionalisation of EETM for extension workers.

Phase 5 activities: third EETM regional workshop

The third EETM regional workshop was held about a year after the second regional workshop in Bali, Indonesia. The time lapse enabled the EETM development team members to consolidate their ideas and implement the recommended activities for a systematic EETM development. The workshop was organised by the Department of Education, Ministry of Agriculture of China, in collaboration with the CIAD, and supported by FAO. This workshop, held in Beijing, China from 26 to 30 August 1996 was to demonstrate the results of the outputs and share the experiences gained during the implementation of the activities. This five-day workshop was attended by 20 participants, the majority of whom were also EETM network members. Among the major outcomes, the workshop was able to:

- provide an insight and systematic inventory-analysis of EETM development and utilisation processes, results, experiences and activities conducted in the last three years;
- better understand and master the EETM development and utilisation process based on shared experiences and lessons learned from various field implementation results;

- suggest action plans for improving EETM development methodology and utilisation process and institutionalisation strategies;
- strengthen the EET network and establish personal contacts for future collaborative activities;
- demonstrate the importance of a systematic approach to training module and curriculum development, the need for networking for effective inter-agency coordination multi-disciplinary approach.

The demonstration of results and sharing of experiences gained from the preparation of the EETM by the network members during the various regional meetings significantly enriched the whole developmental process. As a result of these lessons learned, the best practices were identified for EETM development.

Collaborative Curriculum Development

EET for sustainable agriculture covers a wide range of environmental issues and concerns which should be reflected in the curriculum. In the context of module development, the concept of curriculum and curriculum development may be better understood from the interpretations of the following two definitions:

> A curriculum is a structured set of teaching/learning activities designed to accomplish a set of intended learning outcomes (ILOs) in knowledge, attitude and skill/practice for a specific target audience and context (subject matter, environment, resources, facilities).

> A curriculum is a grand design which spells out the contents to be covered in training, specifies expectations for trainees, delineates procedures for covering contents, suggests the methods for facilitating the learning process, identifies ways for evaluating or assessing learning, and puts everything in a time frame.

In the context of the above definitions, the curriculum development process should be an on-going endeavour that aims to improve the quality of education and training. However, with limited personnel and financial constraints, it may not be practical for a particular agency or institution to undertake this endeavour on a continuous basis or to prepare sets of curricula that will address the training needs on all major environmental issues and concerns. Priorities therefore have to be decided so that the contents of the curriculum/modules are focused only on the most important and critical issues.

Through the collaborative approach of institutional partnership, peer learning and networking, a more efficient utilisation and sharing of limited resources can be effected. Such a collaborative approach has been adopted in the development of the EETM as well as the PETM which clearly demonstrated the improvements of the curriculum development processes, activities and outputs in at least three key areas: (1) enriching the curriculum contents; (2) standard format, structure and quality control; (3) well-tested curriculum development process.

Enriching the curriculum contents

The involvement of at least eight agencies and/or institutions from six different countries not only provided various perspectives and emphases on environmental issues and concerns but also greatly enriched the curriculum contents. For example, the EETM prepared by Thailand's Department of Agricultural Extension (DOAE) focused on the need for environmental conservation to protect the total environment from toxic residues of chemical usage in agriculture. Technical expertise was sought collaboratively with other relevant agencies to enable the DOAE prepare the EETM. The curriculum contents of the EETM prepared by Bangladesh dealt with two key areas: the general aspects of the environment and the protection of agricultural resources. Although the project was undertaken by the GTI, Bangladesh Agricultural University, the main target beneficiaries were the field extension personnel of the DAE, MOA.

The EETM prepared by China and Indonesia both addressed the identification of environmental problems and planning for environmental conservation. China's target beneficiaries included the national-level planners and policy-makers as well as grassroots-level local community leaders. On the other hand, Indonesia targeted only the field-level agricultural extension workers and trainers as its main beneficiaries. China, therefore, prepared two sets of the EETM for the two levels of beneficiaries. The CIAD, China Agricultural University prepared the EETM for the upper level target beneficiaries while the Department of Education (DOE), Ministry of Agriculture, prepared the EETM for the training of grassroots level rural extension workers. This module focused on a practical hands-on approach to agricultural and rural environment conservation and protection in the local community.

In the Philippines, two agencies, the IIRR, and the UPLB, prepared two separate EETMs for a range of target beneficiaries, including extension staff, field workers, teachers and trainers as well as community leaders and farm families. The contents included a wide range of topics such as forest conservation, soil conservation, gender, development and the environment, the greenhouse effect and the agricultural environment, pesticide usage and environmental laws, to cater for the numerous and varied target beneficiaries.

The Malaysian EETM focused on the development of specific knowledge and skills, including the general concept of environment, how to conduct environmental impact assessment (EIA) as well as the preparation of an action plan for an EIA project area. The target beneficiaries were the agricultural field extension workers employed by the DOA and MADA.

Standard format, structure and quality control

As a consequence of the collaborative efforts among the various agencies and/or institutions to develop the EETMs, a standard format and structure was followed. This enabled a more consistent and better quality control process and methods of curriculum development and training implementation. The eight agencies/institutions developed the training modules based on a three-level structure module. These levels were designated (a) the module level, (b) the unit level and (c) the activity level. Training aims and objectives were explicitly stated and the details of the curriculum contents reflected at each level. Such a format and structure thus provided the basis for better quality control of the training activities and outputs as evaluation can be more systematic and objective.

Communication Network and Structure

A communication network consists of interconnected individuals who are linked by patterned flows of information. The study of networks helps illuminate communication structure, the differentiated elements that can be recognised in a system. For instance, the communication structure of a network can reveal which individuals are more or less influential in the system. This communication structure is usually so complex that even in a very small system of individuals, the members of it do not completely understand the communication structure of which they are a part.

REFERENCES

Gruenewald, D.A., 2004, "A Foucauldian analysis of environmental education: toward the socioecological challenge of the Earth Charter", *Curriculum Inquiry* 34(1):71-107.

Malone, K. 1999, "Environmental education researchers as environmental activists", *Environmental Education Research* 5(2):163-177.

Palmer, J.A., 1998, *Environmental Education in the 21st Century: Theory, Practice, Progress, and Promise*, Routledge.

Science (ed.), 1997, "Overhauling environmental education", *Science*, 276:361.

Smyth, J.C. 2006, "Environment and education: a view of a changing scene", *Environmental Education Research* 12(3,4):247-264.

Roth, Charles E. "Off the Merry-Go-Round and on to the Escalator". In *From Ought to Action in Environmental Education*, edited by William B. Stapp, pp. 12-23. Columbus, OH: SMEAC Information Reference Center, 1978. Ed 159 046.

Bibliography

Altieri, Miguel A. 1995. *Agroecology: The science of sustainable agriculture*. Westview Press, Boulder, CO.

Andow, D.A. 1991. "Vegetational diversity and arthropod population response". *Annual Review of Entomology* 36: 561-586.

Anne R. Leslie and Gerrit W. Cuperus (eds.) *Successful Implementation of Integrated Pest Management for Agricultural Crops*. Lewis Publishers, Boca Raton, Florida.

Berry, Wendell. 1996. *The Unsettling of America: Culture and Agriculture*. 3rd edition. University of California Press, Davis. 256 p.

Bird, Elizabeth Ann R., Gordon L. Bultena, and John C. Gardner (eds.) 1995. *Planting the Future: Developing an Agriculture that Sustains Land and Community*. Iowa State University Press, Ames, IA.

Burleson, Wayne, and Connie Burleson. 1994. *Rut Buster: A Visual Goal Setting Book*. Sloping Acre Publishing Company. Absarokee, Montana. 45 p.

Carrol, C. R., J.H. Vandermeer and P.M. Rosset. 1990. *Agroecology*. McGraw Hill Publishing Company, New York.

Dore, J. 1997. *Sustainability Indicators for Agriculture: Introductory Guide to Regional/National and On-farm Indicators*, Rural Industries Research and Development Corporation, Australia.

Finch, C.V. and C.W. Sharp. 1976 *Cover Crops in California Orchards and Vineyards*. USDA Soil Conservation Service, Washington, D.C.

Gliessman, S.R. 1998. *Agroecology: ecological processes in sustainable agriculture*. Ann Arbor Press, Michigan.

Horne, James E. and Maura McDermott. 2001. *The Next Green Revolution: Essential Steps to a Healthy, Sustainable Agriculture*. Food Products Press, an imprint of The Haworth Press, Binghamton, NY. 312 p.

Hudson, Berman E. 1994. "Soil organic matter and available water capacity". *Journal of Soil and Water Conservation*. Volume 49, Number 2. p. 189-194.

Jackson, Wes. 1985. *New Roots for Agriculture*. 2nd edition. University of Nebraska Press, Lincoln, NE. 150 p.

Liebman, M. and T. Ohno. 1998. "Crop rotation and legume residue effects on weed emergence and growth: implications for weed management". In J.L. Hotfield and B.A. Stwerrt (eds.). *Integrated weed and soil management*. Ann Arbor Press, MI

Magdoff, F.R. 1992. *Building soils for better crops: organic matter management*. University of Nebraska Press, NE.

Nair, P.K.R. 1982. *Soil Productivity Aspects of Agroforestry*. ICRAF, Nairobi.

Netting, Robert McC. 1993. *Smallholders, Householders: Farm Families and the Ecology of Intensive, Sustainable Agriculture*. Stanford Univ. Press, Palo Alto.

Pearson, C.J. and R.L. Ison. 1987. *Agronomy of grassland systems*. Cambridge University Press, Cambridge.

Petry, F. 1995*a*, *Sustainability Issues in Agricultural and Rural Development, Vol 1, Trainee's Reader,* Training Materials for Agricultural Planning 38/1, FAO, Rome.

Pretty, J.N. 1994. *Regenerating Agriculture.* Earthscan Publications Ltd., London.

Reijntjes, C.B., Haverkort and A. Waters-Bayer. 1992. *Farming for the future.* MacMillan Press Ltd., London.

Savory, Allan, with Jody Butterfield. 1999. *Holistic Management.* Island Press. Washington, D.C. 616 p.

Sumner, D.R. 1982. "Crop rotation and plant productivity". In M. Recheigl, (ed). CRC *Handbook of Agricultural Productivity,* Vol. I CRC Press, Florida.

Suthipradit, S. et.al. (eds) "Sustainable Agriculture: Possibility and Direction" Proceedings of the 2nd Asia-Pacific Conference on Sustainable Agriculture 18-20 October 1999, Phitsanulok, Thailand.

Thies, C. and T. Tscharntke. 1999. "Landscape structure, and biological control in agroecosystems". *Science* 285: 893-895.

Thompson, Dick. 1997. *Alternatives in Agriculture*: 1996 Report. Thompson On-Farm Research and the Wallace Institute. p. 3-4.

Troeh, F.R., J.A. Hobbs, and R.L. Donahue, et al. 1991. *Soil and Water Conservation.* Prentice-Hall, Englewood Cliffs, New Jersey.

Vandermeer, J. 1989. *The ecology of intercropping.* Cambridge University Press, Cambridge.

Winrock International 1992. *Animal Agriculture on Developing Countries: Technology Dimensions.* Morilton, Arkansas.

World Bank. 1985. *The Small Holder Dimension of Livestock Development: A review of Bank Experience,* Report No. 5979, Operations Evaluation Department, 1985.

Index